Traffic Science

Traffic Science

Edited by

Denos C. Gazis
International Business Machines Corporation
Thomas J. Watson Research Center
Yorktown, New York

A Wiley-Interscience Publication
JOHN WILEY & SONS
New York · London · Sydney · Toronto

242383

Library of Congress Cataloging in Publication Data

Main entry under title:
Traffic science.

  "A Wiley-Interscience publication."
  Includes bibliographical references.
  1. Traffic engineering—Mathematical models—Addresses, essays, lectures. 2. Traffic flow—Mathematical models—Addresses, essays, lectures. I. Gazis, Denos C., 1930–   ed.

HE333.T72   387.3′1   73-21947
ISBN 0-471-29480-2

Printed in the United States of America

10 9 8 7 6 5 4 3 2 1

Contributors

Leslie C. Edie
Research and Development Division
The Port Authority of New York and New Jersey
111 Eighth Avenue
New York, New York

Denos C. Gazis
Department of General Sciences
IBM Thomas J. Watson Research Center
P.O. Box 218
Yorktown Heights, New York

Walter Helly
Department of Operations Research
The Polytechnic Institute of Brooklyn
Brooklyn, New York

Donald R. McNeil
Department of Statistics
Princeton University
Princeton, New Jersey

George H. Weiss
Division of Computer Research & Technology
National Institutes of Health
Bethesda, Maryland

Preface

Traffic started as a blessing and ended up as something between a mixed blessing and a disaster. Few will deny that the automobile satisfied a basic human need for individual mobility. Unfortunately, competition for mobility among individuals over a finite space reduces the mobility of all. Nevertheless, transportation by automobile continues to be an indispensable ingredient of the economies of developed nations and an aspiration of most of the people of developing ones. The problems associated with automobile traffic cannot be wished away in the foreseeable future, but rather methods must be found for dealing with them effectively.

When traffic began to become a problem, it also began to be measured, recorded, and manipulated by more or less empirical methods in an attempt to preserve some semblance of order in the growing plethora of cars. Traffic engineering emerged first as an art and later became increasingly dependent on the same scientific approach that has contributed to the advancement of all other engineering practices. The use of the scientific method in traffic studies has increased significantly over the past fifteen years. Models of traffic flow have been presented and tested against experiments and observations. Other mathematical models have been used for the improved synthesis and operation of traffic systems, using computers for off-line design or on-line adaptive control. This book gives a comprehensive review of these advances in traffic science.

Chapter 1 covers flow theories. It begins with the definition of basic traffic parameters and a discussion of methods for their measurement. There is then a discussion of various models of traffic flow, including both macroscopic continuum models and microscopic car-following models, as well as models that view traffic as a statistical "ensemble" and treat it by techniques of statistical mechanics. The mathematical techniques used in this chapter are calculus and ordinary and partial differential equations. A comprehensive discussion of experiments and observations used to validate various models is also included.

Chapter 2 discusses the delay of traffic at isolated intersections. Traffic streams are described here in terms of probability density functions of the time-gaps (headways) between cars crossing the intersections. Methods of probability theory are then used to estimate such quantities as the average waiting time of an individual car whose movement through the intersection is impeded by cross traffic. In the case of a signalized intersection, the treatment leads to an evaluation of the optimum selection of the cycle and allocation of green time (split) of a traffic light for minimizing total delay to the users of the intersection.

Chapter 3 contains a formulation and discussion of various control problems of traffic in an urban area. Particular emphasis is given to the problem of synchronizing a system of traffic signals of an urban street network, the problem of minimizing delays in congested transportation systems, and the problem of optimizing the operation of critical traffic links such as freeways, tunnels, and bridges. The mathematical techniques employed in this chapter include calculus, modern control theory, mathematical programming, as well as some *ad hoc* schemes for solving problems of a combinatoric nature.

Finally, Chapter 4 contains a discussion of the processes of traffic generation, distribution, and assignment, using techniques of network theory as well as various models which have been suggested to describe the mechanism that creates traffic between pairs of nodes in a network. The chapter ends with a discussion of the planning and management of a transportation network.

This book is intended for research workers in the field and as a reference volume for a graduate or upper division course in traffic theory. The mathematical treatment requires knowledge of calculus, differential equations, and elements of probability and statistics. Certain sections require some familiarity with special mathematical topics. A special effort was made to facilitate the reading of those sections by providing most of the necessary exposition of the underlying mathematical methodology.

Denos C. Gazis

September, 1973

Contents

CHAPTER 1   Flow Theories   1
             Leslie C. Edie

CHAPTER 2   Delay Problems for Isolated Intersections   109
             Donald R. McNeil and George H. Weiss

CHAPTER 3   Traffic Control—Theory and Application   175
             Denos C. Gazis

CHAPTER 4   Traffic Generation, Distribution, and Assignment   241
             Walter Helly

             Index   289

CHAPTER 1

Flow Theories

Leslie C. Edie

··

Contents

I. Introduction   2

     Variables and Parameters   3
     Ranges of Traffic Intensity   4
     Capacity of a Roadway   5
     Bottlenecks   5
     Approaches to Traffic Flow Theory   5
     Experiments   7

II. Definitions, Relationships, and Distributions   7

     Definitions   8
     Relationships   12
     Trajectories   14
     Time Sequence Diagram   18
     Distribution of Traffic Variables   20
        Headway Distributions   21
        Speed Distributions   31

III. Traffic Flow at Different Densities   41

     Light Traffic   42
     Moderate Traffic—Single Lane   44
     Moderate Traffic—Multilane   48
     Heavy Traffic   51

IV. Traffic Dynamics—Macroscopic   51

      Hydrodynamic Theory   52
         Bottlenecks   62
         Mathematical Treatment—Quantitative Models   69
         Some Further Mathematical Considerations   73
      Experiments and Applications of Macroscopic Theory   76
      Summary of Macroscopic Theory   85

V. Microscopic Models   86

      Car-Following Models   86
      Acceleration Noise   92
      Nonlinear Car-Following Models   95
      Other Car-Following Models   98
      Applications of Microscopic Theory   99
      Car-Following and Accidents   101

VI. Recent Work   102

    References   103

I. INTRODUCTION

Traffic flow theories seek to describe in a precise mathematical way the interactions between automobiles, buses, and trucks traveling on roads and highways. In this chapter we shall consider both single and multilane facilities without intersections, weaving sections, ramps, or traffic controls. The behavior of traffic in these environments is discussed elsewhere in this volume. The interaction of a single vehicle with its neighboring vehicles and its changing roadway environment results in a change in direction and/or speed of the vehicle or a continuation of the same course and speed. The alert driver of a vehicle receives information more or less continuously about neighboring vehicles by visual and aural means, he evaluates the information and then may either take prompt action or store information in his memory for possible future action. Similarly, he receives information about the roadway conditions he now encounters or will soon encounter. Also, he receives information about the speed, position, acceleration, and so forth of his own vehicle. His objective generally is to reach a given destination safely at some desired time, plus or minus some margins. If he is early, he may be relaxed and smooth in his actions; if he is late, he may drive in an aggressive, jerky way.

In traffic flow theories, we seek eventually to be able to state some general laws about how a driver and his vehicle respond to a given traffic situation with a given history. Such general laws cannot be expected to have the precision of the laws of physics and astronomy, particularly in dealing with a single driver and his vehicle. However, when the reactions of many driver–vehicle combinations are averaged together, a central tendency can be observed that can be approximated by explicit theories. It is also clear that vehicles must conform to the laws of physics even though drivers, at times, might want to violate those laws to avoid accidents or for other reasons. If a driver's vehicle hits an obstruction traveling at a given speed, the amount of kinetic energy to be dissipated, and therefore the approximate amount of damage, is controlled by physical variables. But, on the other hand, the speed at which a driver will go when unimpeded has little to do with physics unless he attempts to exceed the car's capabilities. It thus seems evident that traffic flow theories will include elements of human psychology and physiology and elements of physics. Until there is some unified traffic flow theory, which can be used to deal with any kind of traffic flow situation, some theories will tend to deal more with the psychological–physiological side, others with the physical side.

Variables and Parameters

Putting this another way, the subject of traffic flow theory is concerned with the behavior, or interactions, of a complex made up of drivers, vehicles, and a roadway, commonly known as the driver–vehicle–roadway complex. However, we are not greatly concerned with a single driver–vehicle combination but with many operating at the same time on the roadway. In fact, one of the principal variables we are concerned with in traffic flow theories is concentration or density measured in vehicles per mile per lane. The others are the traffic flow on a roadway, measured in vehicles per hour per lane; and the traffic speed, measured in miles per hour. It is to be noted that these measurements are values averaged over time, space, and/or a number of vehicles. The driver–vehicle–roadway characteristics enter as parameters into equations dealing with these variables or their inversions. Mean headway, the inverse of flow, is measured usually in seconds; mean spacing, the inverse of concentration, is measured in feet; and mean travel time per unit distance, the inverse of speed, is then measured in seconds per foot. The change of units is made to avoid decimal values. When these variables are correctly defined, the value of any one of them is given by the values of the other two. If q is the flow and k is the concentration, the traffic speed u is given by $u = q/k$. Similarly, if h is the mean headway time between vehicles, and s is the mean spacing rear to rear, the travel time per foot, f, is given by $f = h/s$. These expressions are fundamental to the formulation of traffic flow theories and will be discussed more fully in Section II.

If we are concerned with studying what determines the values assumed by the variables in these expressions and their changes in time and space, it is clear that we must interest ourselves in certain parameters that describe the important characteristics of each part of the driver–vehicle–roadway complex. For the driver, we may be concerned with the parameters of what is called by psychologists the PIEV response to a stimulus, P standing for perception, I for intellection, E for evaluation, and V for volition. The perception involves the reception of visual, aural, and other signals. The intellection involves the recognition or identification of the signals received with similar ones stored in the memory. The evaluation involves an appraisal of the situation and a decision as to what should be done about it. And the volition involves the implementation of the decision by some physical act. In some cases the whole process is treated as a single parameter called the *driver reaction time*.

The vehicle has important parameters in its maximum values of acceleration and deceleration, in its mass, mechanical lags, turning characteristics, and other design features that determine how quickly and in what way it will respond to the driver's commands.

The roadway has important parameters in its lane widths, curvature, grade, smoothness, sight distance, and so on.

Ranges of Traffic Intensity

It is convenient to consider three qualitative ranges of traffic flow intensity when formulating traffic flow theories:

1. *Light traffic*. Vehicles travel at the speeds desired by their drivers, that is, there is little, if any, interference between vehicles, such that a driver is rarely forced to reduce his speed because of the presence of other vehicles on the roadway.
2. *Moderate traffic*. Vehicles form clusters or platoons. This occurs when fast vehicles are forced to slow down temporarily behind slower vehicles. Such impediments are of short duration, and vehicles can achieve average speeds over entire trips that are not far below their desired speeds. They find it relatively easy to pull around the slower vehicles, if the latter do not first move over to a slower lane.
3. *Heavy traffic*. Platoons become very long and tend to run together, forming still larger ones, becoming a continuous stream at certain locations along the roadway. Under these conditions, nearly every driver is constrained to follow the vehicle ahead with little ability to increase his speed by changing lanes. The impediments to free movement are almost continuous, and the journey speed drops very substantially from the desired speed.

There is an analogy that may be made between these qualitative ranges of traffic intensity and some states of physical matter. Light traffic is analogous to the gaseous state, where each molecule is free of attachment to other molecules; moderate traffic is analogous to a vapor state, where some molecules have formed droplets but others are free; and heavy traffic is analogous to a liquid state, where nearly all molecules are in contact with one or more others. Following the analogy further one might add a fourth classification *jammed traffic*, analogous to the solid state of matter, and we can foresee the need for solid-state traffic theory. It could help us know how best to melt the solid mass of traffic and get it back into a liquid state once the traffic stream has congealed.

Capacity of a Roadway

Traffic engineers have long been concerned with determining the capacity of a roadway—that is, the maximum traffic flow that can be sustained over a roadway for a period of an hour or some fraction thereof. The Highway Capacity Manual[1] of the Highway Research Board is very much concerned with this question from both theoretical and empirical points of view. Since traffic flow theory is not yet a well-established science, reliance tends to be placed more on empirical observations than on theory. Data are collected on the maximum observed volumes, that is, flows, found on roadways with similar parameters, and if one is planning a similar roadway, he assumes that the capacity of his roadway will be about that of those presently observed. The traffic scientist, however, is interested in finding out what are the controlling factors that permit and limit the observed maxima.

Bottlenecks

One of the important factors found by traffic scientists is the existence, along a roadway, of bottlenecks—sections that have less capability of accommodating traffic than do other sections of the roadway which are linked in series with the bottleneck sections. It is known that bottlenecks are apt to occur at curves, changes of grade, overpasses, and where the driving task is complicated by accidents or cars parked along the shoulder of the road.

The maximum observed flow of traffic over any substantial period of time upstream or downstream from a bottleneck cannot exceed the capacity of the bottleneck section when there are no intervening ramps.

Approaches to Traffic Flow Theory

The previous paragraphs have given the highlights of the phenomena studied by traffic flow theorists: the variables, parameters, and problems of capacity and bottlenecks. We now consider briefly some of the approaches taken in studying these things. These approaches may be deterministic and deal only with the average values of the variables and parameters, or they may be statistical and deal with the distributions of values as well as their averages. The deterministic approaches may be microscopic and study the typical behavior of one vehicle following another, or they may be macroscopic and study the behavior of sizable groups of vehicles without specifying the behavior of any single vehicle. The microscopic approach has resulted in car-following theories which give formulations for the stimulus–response relationship found when one vehicle seeks to follow behind another vehicle with a lesser desired speed. What the second vehicle does in tracking the trajectory of the first is to regulate its acceleration, considered to be a response to several variables that comprise a stimulus. The most important stimulus is the relative speed—the closing or opening rate—since it determines the sign of the response. If the following vehicle is closing on the vehicle ahead, the response will be to decelerate (negative acceleration); if the following vehicle is falling behind, the response will be positive, that is, to accelerate and catch up. Other stimuli, such as spacing and speed, tend to modulate the magnitude of the response. These are discussed specifically in Section V.

The macroscopic approach to traffic flow is analogous to theories of fluid dynamics or continuum theories. An important and powerful feature of this theory is the conservation of vehicles. It is taken as a physical law that vehicles are neither created nor destroyed in the normal driving process. The theory is not intended to describe the effects of vehicles entering or departing from the traffic stream, but does give valuable insights into the behavior of traffic streams where all vehicles remain on the roadway in question. It is particularly useful in describing the propagation of waves in a traffic stream, their speed of propagation, and the behavior of vehicles passing through the waves.

An important step forward in the development of traffic flow theory was made when a consistency between the macroscopic approach and the microscopic approach was found.[2] By microscopic experiments it has been found possible to predict the parameters of a macroscopic theory.

Statistical theories of traffic flows have been concerned with several aspects. When the traffic intensity is light, it is obvious that statistical theories are more appropriate than deterministic ones since there is little or no interaction between vehicles. If one studies the time distribution of vehicle arrivals at a point on a roadway and assumes that each arrival time is completely independent of any other arrival time and that equal intervals of time are equally likely to contain an equal number of arrivals, it can be shown that the percentage of time intervals having a particular number of arrivals follows the well-known Poisson distribution. This was first pointed out by Adams.[3] If this distribution is found to hold for light traffic, it can be used to describe the distribution of vehicles on a roadway in both space and time.

Another statistical approach to light traffic follows an analogy with the kinetic theory of gases and is concerned with the passing of vehicles on a two-lane, two-way roadway and with the effect of the passing process on their distribution in space and on their speed. In such a theory it is assumed that the process of a fast vehicle overtaking a slower one can be described as a collision of two molecules in which momentum is conserved.

Another type of statistical approach, employed by Prigogine,[4] makes use of statistical mechanics to study the interactions of vehicles. This is known as a Boltzmann-like approach, in which an integral equation for the velocity distribution of cars is studied. For sufficiently small concentrations, it gives the free speed distribution. For finite concentrations, the distribution is altered toward lower velocities. At some critical concentration, the velocity distribution changes abruptly. It is suggested that this change corresponds to going from individual to collective flow conditions.

Also, some researchers employ statistical studies of distributions of the traffic flow variables, principally speed, headway, and counting distributions. Some of the classical theoretical statistical distributions are found to give reasonable representations of these variables under certain roadway and traffic conditions. Among the distributions that have been found applicable are the Gaussian, Pearson, log–normal, exponential, Erlang, gamma, Poisson, and various mixed distributions. These are discussed in Section III.

Experiments

In the final part of this chapter, we shall discuss some of the important experimental work that has been conducted in the process of testing and formulating theories. Traffic science is very much experimental, as well as theoretical. At the present stage of its development, the theoretical side of traffic science is much further advanced than the experimental side, but significant advancements have been made in the equipment available for instrumenting vehicles and roadways, and for making observations by using photography instead of instruments.

II. DEFINITIONS, RELATIONSHIPS, AND DISTRIBUTIONS

This section discusses the general properties of road traffic and the basic relationships that provide a background of information describing the context within which traffic flow theories are formulated. As in any science, one important step to take is the definition of variables and parameters in precise terms that are generally acceptable to most researchers. This promotes communication and understanding between various researchers.

Definitions

The definitions of traffic flow, concentration, and speed that we give here are operational, that is, they are based on the ways in which these variables are observed and measured. They are also consistent with the relationships $q = uk$, where the flow q is directly proportional to the product of the speed u and the concentration k, and by an appropriate choice of units the constant of proportionality can be made unity. The significance of this relationship can be readily understood by means of an analogy to hydraulics, wherein we choose units as follows: We measure the flow q of water in a pipe in gallons per minute, the mean speed of flow in feet per minute, and the concentration k in gallons per foot. It is evident that in a minute uk gallons of water would be discharged. With traffic as with fluids it is relatively easy to measure the flow past a point and to calculate or measure the concentration. It is not, in either case, easy to measure the mean speed since molecules and vehicles can be observed to move at different speeds at different times and locations, and it is not self-evident how these speeds should be averaged so that the $q = uk$ relationship holds.

Wardrop[5] first showed how to determine the mean speed of a traffic stream by arguments similar to the following: In a steady-state condition we observe k_i vehicles/mile traveling at a speed of u_i miles/hr at any point along a roadway. These vehicles contribute to the total flow an amount $q_i = u_i k_i$ vehicles/hr. The total flow is then $q = \sum u_i \cdot k_i$. The total concentration is $k = \sum k_i$. Therefore, the mean speed must be $u = q/k = \sum u_i k_i / \sum k_i$. In this we are weighting the speed classes u_i by the number of vehicles per mile having each speed class. Since this is weighting according to the number of vehicles found in a unit space, it is termed *the space mean speed*. We may eliminate k_i from the expression for u by substituting $k_i = q_i/u_i$. We then obtain the expression $u = \sum q_i / (\sum q_i/u_i)$. Since q_i is the number of vehicles passing a point along the roadway in a unit of time, we find that the mean stream speed is given by the harmonic (rather than the arithmetic) mean when we use the speed distribution of vehicles passing a point along the roadway.

Now, traffic is observed in basically two ways. First, observations may be made at two or more fixed points located along a roadway where measurements are taken of the time of occurrence of traffic events (or the intervals of time between them) at each of the points. Traffic events for this purpose are defined as the arrival of some part of a vehicle, usually the front or the rear or both. Second, observations may be made at two or more instants of time. The events observed at each instant are the positions of each vehicle on the roadway. The first way uses a short increment of roadway, dx, and a large interval of time T; the second uses short intervals of time, dt, and a large section of roadway X. This is illustrated in Figure 1, where the trajectories of some 25 vehicles in space X and time T are shown as moving points. The point taken may be either the front or, preferably, the rear.

If $N = \sum n_i$ is the total number of vehicles traversing dx in time T with n_i at speed u_i, we can compute the values of the three traffic variables as follows, remembering that vehicles distributed in time have an average speed based on the harmonic average:

$$q = \frac{N}{T},$$

$$u = \frac{N}{\sum n_i/u_i},$$

$$k = \frac{q}{u} = \frac{\sum n_i/u_i}{T}.$$

By substituting $u_i = dx/t_i$, where t_i is the travel time across dx, and by multiplying the expression for q by dx in numerator and denominator, we obtain

$$q = \frac{N\,dx}{T\,dx},$$

$$u = \frac{N\,dx}{\sum n_i t_i}, \tag{1}$$

$$k = \frac{\sum n_i t_i}{T\,dx}.$$

For the second way of measuring, we let $N = \sum n_i$ be the total number of vehicles observed on a stretch of length X. Again we can compute the values of the variables, but in this case we remember that the mean speed is the space mean. Results are the following:

$$k = \frac{N}{X},$$

$$u = \frac{\sum n_i u_i}{N},$$

$$q = uk = \frac{\sum n_i u_i}{X}.$$

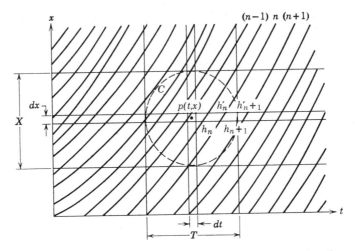

FIGURE 1. Vehicle space–time trajectories and methods of measurement.

Again substituting for u_i, which in this case is x_i/dt, where x_i is the distance traveled by vehicles at speed u_i in a time interval dt, we obtain

$$k = \frac{N\,dt}{X\,dt},$$

$$u = \frac{\sum n_i x_i}{N\,dt}, \qquad (2)$$

$$q = \frac{\sum n_i x_i}{X\,dt}.$$

A comparison of Eqs. 1 with Eqs. 2 shows that they are equivalent and lead to the following general definitions:

1. The mean flow q of vehicles traveling over a roadway of length X during a time T is the total distance traveled on the roadway by all vehicles which were on it during any part of time T divided by XT, the area of the space–time domain observed.
2. The mean concentration k of vehicles traveling over a roadway of length X during a time T is the total time spent by all the vehicles on X during time T divided by XT.
3. The mean speed u of vehicles traveling over a roadway of length X during time T is the total distance traveled on the roadway by all vehicles that were on it for any part of time T divided by the total time spent by all vehicles on the roadway during time T.

These definitions may be further generalized by using the area of any shape of space–time domain. Thus if one wishes to find the mean value of the variables at a point (x, t), he may use a circular or elliptical space–time domain, as well as the square or rectangle shown in Figure 1.

A further generalization can be made, as pointed out recently.[94] With reference to Figure 6, one observes that the triangularly shaped area $OP'N'$ on the N–x plane is a projection of the $N(x, t)$ surface OPN on that plane, and OQN is its projection on the x–t plane. It will be noted that the area of $OP'N'$ is

$$A_t = \tfrac{1}{2}(x_o + x_N) + \sum_{i=1}^{N-1} x_i,$$

the sum of the x_i in the space–time domain OQN which can be called A_n. Similarly, $OPQ = A_x$ is the projection of the surface on the N–t plane. Thus we can generalize:

$$q = \frac{A_t}{A_n},$$

$$k = \frac{A_x}{A_n},$$

$$u = \frac{A_t}{A_x}.$$

The variables q and k are by their nature averages over a number of vehicles since they count the number of vehicles appearing in a unit of time and in a unit of space, respectively. This is not necessarily true of their inverses, headway h and spacing s, which can relate to a single vehicle following another vehicle. In measuring spacing it seems preferable to measure from the rear of the following vehicle to the rear of the leading vehicle since this gives the space occupied by the follower's vehicle in addition to the clear space which is under his control. The headway time between them is not exclusively associated with the follower since it depends on the speed set by the leading vehicle as well as the spacing selected by the follower and on the functional relationship between the two.

Occupancy is a traffic term that has recently been found useful in traffic research involving field observations because it can be measured by simple instruments. It is a nondimensional variable giving the proportion of a roadway length occupied by vehicles. It can be measured on a single aerial photograph, or it can be measured by observing the proportion of time a single point on a roadway is covered by vehicles. If we accept the aerial photo as giving the true occupancy, then the observation at a point gives the same value when the traffic stream is in a steady-state condition. We define the steady-state condition as one where the joint distribution of vehicle speeds and lengths is the same at all points along the roadway.

If on the average there are n_i vehicles of length L_i on the roadway, they occupy space of $n_i L_i$; and if the roadway is one lane and has unit length, the space is $k_i L_i$. For example, if there are 10 vehicles/mile, 16 ft in length (0.003 mile), the occupancy is 0.03; 3% of the mile length of lane is covered by 16-ft vehicles. Thus the occupancy of the ith class is $C_i = k_i L_i$, and the total occupancy of all classes is

$$C = \sum_i k_i L_i.$$

The n_i vehicles all of length L_i will be observed to have various speeds u_j, where $j = 1, 2, 3, \ldots, n_i$ is the order of their arrivals at the observation point. The time the L_i vehicles cover the point on the roadway being observed is

$$t_i = L_i \sum_j \frac{1}{u_j}.$$

The time the point is covered by any vehicle is

$$\sum_i t_i = \sum_i L_i \sum_j \frac{1}{u_j},$$

and if L_i is in miles and u_j in miles per hour, t_i is in hours. We note that the second factor, $\sum_j (1/u_j)$, is related to the mean speed of the L_i vehicles, u_i, as follows:

$$\sum_{j=1}^{n_i} \frac{1}{u_j} = \frac{n_i}{u_i}.$$

The time for n_i vehicles to pass the point is $T = n_i/q_i$. Thus the proportion of an hour during which the point is covered by any vehicles is given by

$$C = \sum_i \frac{t_i}{T} = \frac{\left(\sum_i L_i n_i / u_i \right)}{(n_i/q_i)},$$

$$C = \sum_i L_i k_i.$$

Thus a traffic stream in steady state occupies a point for a proportion of time equal to the proportion of roadway covered by the vehicles at a point in time.

Relationships

One of the first relationships investigated among traffic variables was that between spacing and speed. It is assumed that spacing is a function of speed, that is, $s = f(u)$. One of the first theories was that the spacing should be sufficient to avoid a rear-end collision when the vehicle ahead stopped instantaneously. Three terms are required to take care of this, the vehicle length, the distance traveled during the response time before the second vehicle could achieve maximum braking, and the distance traveled during deceleration. This has the quadratic form

$$s = L + bu + cu^2, \tag{3}$$

where

$s = $ minimum safe spacing,

$L = $ vehicle length,

$b = $ response time,

$c = \dfrac{1}{2a}$, where $a = $ maximum deceleration.

Typical values might be $L = 15$ ft, $b = 1$ sec, and $c = \frac{1}{40}$ sec^2/ft ($a = 20$ ft/sec^2). This gives $s = 95$ ft at $u = 40$ ft/sec, yielding a flow rate of $q = (40/95)3600 \simeq 1500$ vehicles/hr at that speed. Other speeds give different flows. It is known that much greater flows than this are common, and therefore, this functional is conservative, especially considering that many drivers will drive at more than the minimum safe spacing. It was, therefore, suggested that the last term be dropped on the basis that the lead car would not stop instantaneously, but only at maximum deceleration and that the two cars would have the same maxima. This would reduce the spacing to 55 ft and yield a flow of $(40/55)3600 \simeq 2600$ vehicles/hr at $u = 40$ ft/sec. This is close to the maximum observed flows on any roadway at any speed and is therefore too liberal. This formula describes the well-known rule taken from the California Vehicle Code, which prescribes one vehicle length (about 15 ft) for each 10 miles/hr of speed. In this case $b = 1/14.5 = 1.02$ sec for a 15-ft vehicle length.

One reason for studying such functional relationships is to determine the limiting capacity of a roadway and the speed at which this limit is reached. This speed has been called the optimal traffic speed u_m, and it is the speed at maximum flow. For the assumptions above, we found flows of 1500 vehicles/hr and 2600 vehicles/hr at $u = 40$ ft/sec. To find the maximum flow, we find the minimum of the average headway $h = s/u$ by differentiation:

$$h = \frac{L}{u} + b + cu,$$

$$\frac{dh}{du} = -\frac{L}{u^2} + c.$$

Setting it to zero, we find

$$u = \sqrt{L/c}.$$

Substituting assumed values for the first case, we find

$$u = \sqrt{15 \times 40} = 24.5 \text{ ft/sec}.$$

The spacing at this speed is

$$s = 15 + 24.5 + \frac{(24.5)^2}{40} = 54.5 \text{ ft}.$$

The maximum flow is

$$q_m = 3600 \frac{u}{s} = 3600 \frac{24.5}{54.5} = 1620 \text{ vehicles/hr}.$$

For the second case, where $c = 0$ and s is linear in u, the maximum flow occurs when u is infinite; but since this is not possible, maximum flow would occur at the maximum speed attained, somewhere around the speed limit or the mean free speed. If this is 60 miles/hr, 88 ft/sec, the corresponding mean spacing would be

$$s = L + bu = 15 + 88 = 103 \text{ ft},$$

and the maximum flow would be

$$q_m = 3600 \frac{u}{s} = 3600 \frac{88}{103} = 3076 \text{ vehicles/hr}.$$

If the speed could approach infinity, we would have

$$q_m \rightarrow 3600 \text{ vehicles/hr},$$

and the headway time would be $h = b$, the response time.

Since actual maximum flows are between the values computed in the two cases, it is evident that either none gives the correct formulation or the values of the parameters used are incorrect.

Trajectories

One of the most effective forms in which to study traffic behavior is the space–time diagram showing the trajectories of each vehicle in both space and time. Figure 2 shows the difference in s and h for vehicles following the quadratic-type spacing function at 60 ft/sec, 10 ft/sec, and the optimal 24.5 ft/sec.

u	s	h	q
60	165	2.75	1310
24.5	54.5	2.22	1620
10	27.5	2.75	1310

In Figure 2, vehicle 1 is traveling at $u = 60$ ft/sec and vehicle 2 is following at spacing of $s = 165$ ft and a headway of $h = 2.75$ sec. Vehicle 3 is going at $u = 24.5$ ft/sec, and 4 follows at $s = 54.5$ ft, $h = 2.22$ sec, the optimum for the given quadratic function. Vehicle 5 is going at $u = 10$ ft/sec, and 6 follows at $s = 27.5$ ft, $h = 2.75$ sec.

In Figure 1, 25 vehicles are shown on the space–time diagram. They are not in a steady-state condition, although those in the strip dx appear approximately that way. We observe that vehicle headways and spacings are increasing, and vehicles are accelerating because their trajectories are increasing in slope.

The short length of roadway dx represents what is commonly called a "trap." With detectors at each end of dx, the speeds and headways can be readily observed. Such setups were once employed by traffic police to "trap" speeders before the advent of radar. Speed is determined by the length of dx

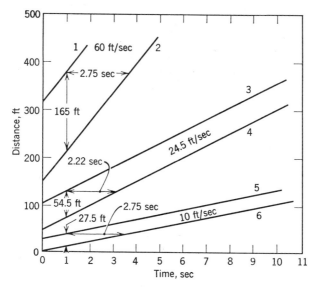

FIGURE 2. Space–time trajectories showing the spacing and headways between vehicles at three speeds.

divided by the transit time. Headways are given by the difference in arrival times at the entering point of the trap, h_n and h_{n+1}, or at the existing point, h'_n and h'_{n+1}. This method of observation is still used by traffic experimenters to measure flow variables, as described under Definitions (see Eq. 1), when the space–time domain has an area of $T\,dx$. The values of q, u, and k so measured would be those associated with a centroid at a point $x = dx/2$, $t = T/2$, the center of the rectangle.

The other principal method of observation is represented by the rectangle $X\,dt$. Data prescribing the trajectories in $X\,dt$ would be obtainable from two photographs made time $t = 0$ and $t = dt$ of a roadway 0–X. The values of k, u, and q so obtained would represent the average around a point $x = X/2$, $t = dt/2$.

By repeating detectors at intervals of dx throughout X, or by taking photos at intervals of dt throughout T, the trajectories in the rectangle XT can be established, and averages can be taken over XT or a portion of it as within the circle C. Such averages would be associated with the point $P(x, t)$ at the centroid of the areas. Finding such averages is useful in studying traffic behavior using a fluid analogy, as is explained in Section IV (Traffic Dynamics—Macroscopic).

Figure 3 shows details of trajectories crossing a strip dx when four events are considered, as indicated for the $(n - 1)$st vehicle: $t_{A1}(n - 1)$, arrival time of the front of the vehicle at the first point $(x = 0)$; $t_{A2}(n - 1)$, arrival of the front at the second point $(x = dx)$; $t_{D1}(n - 1)$, the departure of the vehicle from point 1; and $t_{D2}(n - 1)$, the departure from point 2. One may then set up equations in some of the intervals between these events for two vehicles, the $(n - 1)$st and the nth, and calculate the microscopic variables for the nth vehicle: L_n, length; u_n, speed; a_n, acceleration; h_n, headway; and s_n, spacing. The equations are

$$L_n = u_n t_2 - \tfrac{1}{2} a_n t_2^2, \tag{4}$$

$$dx = u_n t_1 + \tfrac{1}{2} a_n t_1^2, \tag{5}$$

$$dx = (u_n - a_n t_2) t_3 + \tfrac{1}{2} a_n t_3^2, \tag{6}$$

$$s_n = u_n h_n - \tfrac{1}{2} a_n h_n^2, \tag{7}$$

where

$$t_1 = t_{D2} - t_{D1},$$

$$t_2 = t_{D1} - t_{A1},$$

$$t_3 = t_{A2} - t_{A1},$$

$$h_n = t_{D1}(n) - t_{D1}(n - 1),$$

and it is assumed that the acceleration a_n remains constant.

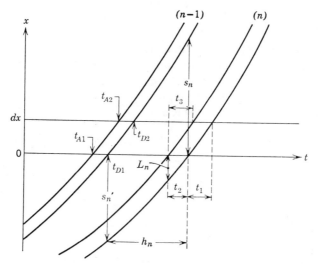

FIGURE 3. Space–time trajectories showing measurements on a small increment of roadway dx.

Solving for the variables of interest, we find that

$$a_n = \frac{2t_1 - 2t_3}{t_1 t_3^2 - 2t_1 t_2 t_3 - t_1^2 t_3} \, dx, \tag{8}$$

$$u_n = \frac{t_3^2 - 2t_2 t_3 - t_1^2}{t_1 t_3^2 - 2t_1 t_2 t_3 - t_1^2 t_3} \, dx, \tag{9}$$

$$L_n = \frac{t_2 t_3^2 - t_1 t_2^2 - t_2^2 t_3 - t_1^2 t_2}{t_1 t_3^2 - 2t_1 t_2 t_3 - t_1^2 t_3} \, dx, \tag{10}$$

$$S_n = \frac{h_n t_3^2 - 2h_n t_2 t_3 - h_n t_1^2 - t_1 h_n^2 + t_3 h_n^2}{t_1 t_3^2 - 2t_1 t_2 t_3 - t_1^2 t_3} \, dx, \tag{11}$$

when

$$a_n = 0, \qquad t_1 = t_3,$$

and

$$u_n = \frac{dx}{t_1},$$

$$L_n = \frac{t_2 \, dx}{t_3},$$

$$S_n = \frac{h_n \, dx}{t_1} = h_n u_n.$$

Time Sequence Diagram

Another useful form in which to study traffic behavior is the time sequence diagram, in which we diagram the cumulative number vehicle N passing a point versus time. Figure 4 gives such a chart. In it the 0th vehicle passed at $t = 0$, and we observe that the ith vehicle passed at $t = t_i$.

The headway time for the ith vehicle is

$$h_i = t_i - t_{i-1},$$

and the total time for n vehicles to pass is

$$\Delta t = t_{i+n} - t_i.$$

The flow generated by the n vehicle is

$$q = \frac{n}{T} = \frac{\Delta N}{\Delta t},$$

which is the slope of the $N(t)$ curve, and if we treat the stream of vehicles as a continuum (by some averaging process), we can find the flow at a point x and time t as

$$q_{x,t} = \lim_{\Delta t \to 0} \frac{\Delta N_x}{\Delta t} = \frac{dN_x}{dt}.$$

If observations are made at two points, we have two curves $N_1(t)$ at point X_1, and $N_2(t)$ at point X_2, as shown in Figure 5. The horizontal distance

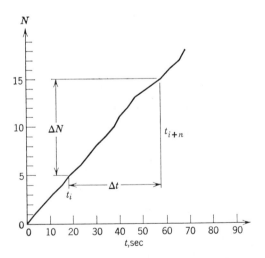

FIGURE 4. Vehicle number N versus time t.

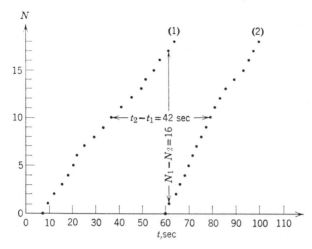

FIGURE 5. Vehicle number versus time diagram for two observation points.

between two points on the two curves gives the travel time of the vehicle having the value of N given by the two points. Thus

$$u_n = \frac{X_2 - X_1}{t_2 - t_1} = \frac{\Delta X}{\Delta t} \,,$$

and in the limit as $x \to 0$,

$$k(t) = \frac{dN}{dx} \,.$$

It is obvious that we can similarly construct curves giving $N(x)$ versus x by observing at points in time instead of points in space. The slopes of such curves then give limiting values of k.

It is more interesting to consider a three-dimensional chart that would show the positions of each vehicle simultaneously in both space and time. Such a diagram is illustrated in Figure 6. Here are shown the trajectories of vehicles 0–6 inclusive in three dimensions, N, x, t. Successive vehicles at first slow down and then speed up, vehicle 6 traveling at approximately the same speed as vehicle 0.

The vehicle trajectories form a curved (or broken) surface. Intersections of this surface with several plane surfaces are shown, namely, $X = 0, 200, 400$, and 600 ft, and $t = 20$ sec.

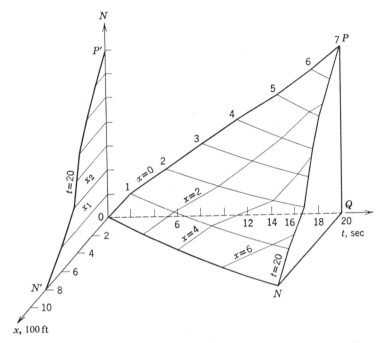

FIGURE 6. Vehicle number versus space and time showing trajectories of several vehicles in the three dimensions.

Distribution of Traffic Variables

In the previous discussion the traffic variables q and k have been considered as averages over a number of vehicles, and h and s, their inversions, as averages or as values observed for a single vehicle; speed has also been considered both ways. We now recognize the fact that not only are the averages and individual values important to traffic flow theories, but also the distributions of the individual values are important. Traffic flow is a very complex phenomenon, and one way of simplifying the task of studying it is to break the traffic stream down into its component parts and study these in isolation. In this way the number of variables is reduced and the likelihood of obtaining insights into traffic behavior may be increased. Some theorists have felt that the distributions of headways, speeds, and spacing are sufficiently important to be studied in their relative isolation.

Mathematical statistics and probability theory provide the tools with which distributions may be studied without knowing the underlying complex phenomena which produce the distributions. When data are found to fit a theoretical distribution, there is an implication that the assumptions of the theory are appropriate *statistical* assumptions for the process. For example, if traffic data are found to conform to a binomial or a Poisson distribution, the implication is that the traffic events measured, for example, the inter-arrival times of successive vehicles, occur independently, that is, there is no significant underlying situation deviating from a basic probabilistic occurrence of events.

In addition to giving insight into the nature of an underlying process, theoretical distributions permit large amounts of data to be summarized into a few parameters that can become part of a stochastic theory of traffic flow.

Headway Distributions

We can find the distribution of headways by observing the times of arrival of successive vehicles in a given lane at a point along the lane. As noted previously, we define the headway of the ith vehicle as

$$h_i = t_{i-1} - t_i,$$

where t_{i-1} is the arrival time of the vehicle ahead and t_i that of the ith vehicle. If we observe many vehicles, we can obtain the distribution of h_i and investigate whether this variable can be described by some theoretical distribution function.

Some of the characteristics of such distributions are well established. For example, under light traffic conditions, the headway distribution function reduces to a negative exponential or displaced negative exponential distribution. These functions describe the distribution of certain random events, such as the intervals between radioactive atomic fissions observed on a Geiger counter.

It was first suggested by Adams[3] that the number of vehicles in light traffic passing a point in equal intervals of time follow a Poisson distribution. If this is the case, the distribution of headways can be shown to be described by the negative exponential distribution. The Poisson distribution gives the probability, or proportion of a number of equal time intervals, during which any number of vehicles, n, will arrive at a point as

$$P_n(t) = \frac{e^{-qt}(qt)^n}{n!},$$

where qt is the mean number of vehicles arriving during a time interval of, say, t sec. Thus

$$P_0(t) = \text{probability of no vehicle in } t \text{ sec} = e^{-qt},$$

$$P_1(t) = \text{probability of 1 vehicle in } t \text{ sec} = qte^{-qt},$$

$$P_2(t) = \text{probability of 2 vehicles in } t \text{ sec} = \tfrac{1}{2}(qt)^2 e^{-qt},$$

$$P_3(t) = \text{probability of 3 vehicles in } t \text{ sec} = \tfrac{1}{6}(qt)^3 e^{-qt}.$$

With the Poisson counting distribution there is no upper limit on how many vehicles might arrive in a given time interval. Sometimes this is not the case. For example, suppose a parking lot with N-vehicle capacity empties in a time T. If any vehicle selected at random has a probability p of leaving in any small time interval, the counting distribution can be given by the binomial expression:

$$P(n) = C(N, n)p^n(1 - p)^{N-n},$$

where

$P(n) = $ probability that n vehicles leave in a small time interval Δt, if

$p = \Delta t/T = $ probability that any vehicle leaves in a particular interval Δt,

$N = $ number of vehicles leaving during T,

$t = $ any small part of time T,

$C(N, n) = $ combination of N things taken n at a time,

$$= \frac{N!}{n! \, (N - n)!} \, .$$

If $N = 10$, $T = 60$ min, and $\Delta t = 1$ min, then $p = \frac{1}{60}$, the average departure rate would be $m = Np = 0.167$ vehicle/interval, and the variance would be $Np(1 - p) = 0.164$. The probability density would be the following:

$$P(0) = 0.845,$$

$$P(1) = 0.143,$$

$$P(2) = \frac{0.012,}{1.000}$$

and the probability of three or more exiting at once would be small, less than once in 20 hr.

When N is small, the computations are not tedious. As N gets larger, tables may be used, but even this becomes tedious with increasing values for N. However, if N is very large and p is small, it has been shown by Molina[7] that the Poisson distribution may be used as an increasingly close approximation to the binomial.

Also, when the mean value exceeds a value of about 5, the normal distribution becomes a close approximation. Table 1 shows the fit of both the Poisson and normal density functions to data giving the number of vehicles arriving in $\frac{1}{2}$-min intervals when the rate of flow is 600–700 vehicles/hr for a total of 450 intervals. The mean rate of flow is 5.46 vehicles/interval, and the variance is 2.73 vehicles.

TABLE 1

COMPARISON OF POISSON AND NORMAL COUNTING FUNCTIONS TO
OBSERVED DENSITY

	Observed Occurrences	Frequencies		
		Actual	Poisson	Normal
n		f	f_P	f_N
0	6	1.3	0.4	1.11
1	18	4.0	2.3	2.77
2	35	7.8	6.3	5.71
3	52	11.6	11.5	9.81
4	68	15.1	15.7	14.04
5	71	15.8	17.2	16.75
6	55	12.2	15.6	16.62
7	45	10.0	12.2	13.73
8	39	8.7	8.3	9.45
9	26	5.8	5.1	5.42
10	14	3.1	2.8	2.58
11	8	1.8	1.4	1.0
12	6	1.3	0.6	0.4
13	4	0.9	0.3	0.1
14	2	0.4	0.1	0.02
15	1	0.2	0.04	0.01
	450	100.0	100.0	100.0

$\chi_P^2 = 0.74.$
$\chi_N^2 = 0.46.$

For these data, the normal fits slightly better than the Poisson, but both
give excellent fits.

The probability function for headways when the arrival or counting
function is Poisson can be obtained as follows: We treat the time interval t
as a variable in the Poisson expression for $P(0)$, which gives

$$P_0(t) = e^{-qt} \qquad \text{or} \qquad P_0(t) = e^{-t/h},$$

where h is the mean headway. This function gives the probability that no
vehicle arrives in an interval t. The probability that some vehicle arrives in t
is the same as the probability that a headway h_i is equal or less than t, namely,

$$P(h_i \leq t) = 1 - e^{-t/h}.$$

This is a probability distribution function, and we can obtain the probability density function by differentiation:

$$P'(h_i \leq t) = P(h_i = t) = \frac{1}{h} e^{-t/h}.$$

While the Poisson and negative exponential distributions are very useful in traffic flow theories for light traffic, especially because of their mathematical simplicity, they have shortcomings in not taking into account the limitations imposed by the physical size of vehicles and finite velocities. These limitations make it less realistic for a single lane without passing than for multiple lanes.

One way to deal with the effects of these limitations is to replace the negative exponential headway distribution by its "displaced" analog:

$$P(h_i = t) = 0 \qquad\qquad 0 \leq t \leq t_0$$

$$= \left(\frac{1}{h}\right) e^{-t/h} \qquad t_0 \leq t,$$

$$P(h_i \leq t) = 0 \qquad\qquad 0 \leq t \leq t_0$$

$$= 1 - e^{-t/h} \qquad t_0 \leq t,$$

where t_0 is the minimum allowable headway. This is equivalent to saying the following:

The probability that a headway, starting at $t = 0$, ends in the interval $(t, t + dt)$, provided that it did not end before time t is

$$0 \quad\text{if}\quad (t < t_0)$$

or

$$1/h \quad\text{if}\quad (t_0 \leq t).$$

The counting distribution for the displaced exponential headway distribution has been derived by Oliver.[8] It is given by the following equations:

$$P_n(t) = 0 \qquad\qquad 0 \leq t < nt_0$$

$$= \frac{\gamma[n; q(t - nt_0)]}{(n - 1)!} \qquad nt_0 \leq t < (n + 1)t_0$$

$$= \frac{\gamma[n; q(t - nt_0)]}{(n - 1)!}$$

$$- \frac{\gamma[n + 1; q(t - (n + 1)t_0)]}{n!} \qquad (n + 1)t_0 \leq t,$$

where $\gamma(n; x) = $ incomplete gamma function of order n and argument x

$$= \int_0^x e^{-t} t^{n-1} \, dt$$

$$= (n-1)! \left[1 - e^{-x} \left(\sum_{m=0}^{n-1} \frac{x^m}{m!} \right) \right] \qquad \text{for} \quad n > 0.$$

The displaced negative exponential distribution has been shown by Oliver to fit actual headway data for streets in San Francisco and Boston.

Counting distributions are affected by the duration of the time interval during which the count is made. It has been a common practice to take short period counts of one or more minutes and to use the maxima of these to estimate roadway capacity. Newell[9] has investigated the validity of this hypothesis versus the contrary hypothesis that the peak short-time values are mere statistical fluctuations.

Suppose we look for the maximum flows occurring in any m consecutive minutes and in any n consecutive minutes where $n > m$. If the average flow in the longer intervals is Z/n, the expected number in m minutes is $(m/n)Z$. If the peaking factor F is the proportion of Z occurring in the maximum group of m minutes, that is, the ratio of the average flow to the maximum flow observed, then

$$\frac{m}{n} \le F \le 1,$$

and F can be expressed in terms of the sum Y_i of the flows X_i occurring in all groups of m minutes, their standard deviation and the value of m. Newell did this and found that for a fixed ratio of n/m the value of F was not very sensitive to the value of m; for example, the difference between $m = 15$ min and $m = 30$ min is less than $\frac{1}{10}$.

Although some observers have believed that peak short-time flows (5 or 6 min) give a better indication of highway capacity than peak flows over longer times (15–30 min), this theory would suggest that any difference found is due mostly to random fluctuations that have no obvious connection with capacity.

Dunne et al.[10] proposed a discrete Markov model that accounts for the bunching tendency in traffic by assuming correlations between successive vehicles. Time interarrival experiments were conducted to test the model.

There are two characteristics of vehicular traffic that distinguish it from, say, telephone traffic. The first is the discrete structure of the traffic stream, consisting of vehicles of finite length, which prevents the occurrence of very small headway in single-lane traffic. The other is the tendency to bunch when passing is hindered on a road with two-way traffic.

If
$$X_k = \text{number of vehicles which pass a point in an interval}$$
$$kt \text{ to } kt + t, k \text{ being any integer}$$

and
$$P(X_{k+1} = 1 \mid X_k = 0) = p,$$
$$P(X_{k+1} = 1 \mid X_k = 1) = 1 - q,$$

the Markov process matrix is

$$
P = \begin{array}{c c} & \begin{array}{c c} 0 & 1 \end{array} \\ \begin{array}{c} 0 \\ 1 \end{array} & \left[\begin{array}{c c} 1-p & p \\ q & 1-q \end{array} \right] \end{array},
$$

neglecting the possibility of two or more vehicles in the time interval t.
  The counting distribution

$$Y_n = X_1 + X_2 + \cdots + X_n$$

has well-known properties; for example, Y_n is asymptotically normal with mean $np/(p + q)$ with variance $npq(2 - p - q)/(p + q)^3$.
  Both platoons and gaps have the geometric distribution

$$P(k) = q(1 - q)^{k-1} \qquad k \geq 1.$$

  Another approach is to consider a traffic stream of moderate intensity, where some vehicles are free-flowing and others are restrained in platoons behind another vehicle. If the proportion of all vehicles in platoons is a, the proportion which is free-flowing is $1 - a$, a composite distribution of headways of all vehicles is given by

$$P(h \geq t) = (1 - a) \exp \left[\frac{-(t - t_1)}{h_1 - t_1} \right] + a \exp \left[\frac{-(t - t_2)}{h_2 - t_2} \right], \qquad (12)$$

where
$$t_1 = \text{minimum headway of free-flowing vehicles,}$$
$$h_1 = \text{mean headway of free-flowing vehicles,}$$
$$t_2 = \text{minimum headway of restrained vehicles,}$$
$$h_2 = \text{mean headway of restrained vehicles.}$$

  This composite distribution has also been used in recent studies of traffic flow on two-lane urban streets.[11,12] Figure 7 shows a cumulative composite distribution fitted to two sets of experimental data A and B.

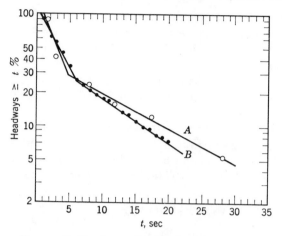

FIGURE 7. Headway distributions for two sets of data fitted by composite negative exponential distributions.

One can observe approximately the following parameters for the two composite negative exponential distributions, as given by Eq. 12:

	A	B
Proportion of restrained vehicles (a)	0.60	0.55
Minimum free-flowing headway (t_1)	5 sec	6 sec
Minimum restrained headway (t_2)	1 sec	$\frac{3}{4}$ sec

Davison and Chimini[13] suggested use of a composite hyperlang distribution, where

$$P(h \geq t) = (1 - a) \exp\left(\frac{-(t - \delta_1)}{\gamma_1 - \delta_1}\right)$$

$$+ a \exp\left(\frac{-k(t - \delta_2)}{\gamma_2 - \delta_2}\right) \sum_{x=0}^{k-1} k\left(\frac{t - \delta_2}{\gamma_2 - \delta_2}\right)^x \frac{1}{x!}$$

where

$(1 - a)$ = the proportion of free headways,

a = the proportion of constrained headways,

δ_1 = minimum free headway,

δ_2 = minimum constrained headway,

γ_1 = mean free headway,

γ_2 = mean constrained headway,

k = an index denoting the degree of nonrandomness in the constrained headway distribution.

In addition to fitting negative exponentials and hyperlangs to a composite distribution of restrained and free-flowing vehicles, other distributions may be used for either group. Such composite distributions may also be called "mixed" distributions.

Buckley[14] has considered the use of delta, gamma, Erlang, Pearson, and Gaussian functions, but they appeared to offer no advantage over a composite negative exponential.

A distribution of particular interest which has been found to fit headway data quite well in tunnels and on bridges as well as on streets and highways is the log–normal distribution, in which the logarithm of the variable is found to fit a Gaussian distribution. Thus the log–normal distribution is

$$P(h) = (h\sigma\sqrt{2\pi})^{-1} \exp\left[-\frac{1}{2\sigma^2} (\log h - m)^2 \right],$$

where m is the mean and σ^2 the variance of $\log h$.

Figure 8 shows the fit of the log–normal distribution to headway data by Buckley[15] on one lane of a freeway and Daou[16] in a long under-river tunnel. These data are plotted on log–probability scales. Variables taken from a log–normal distribution appear as straight lines on these coordinates. The tunnel data show much longer headways than the freeway data, as would be expected, but both show a reasonably good fit to the model.

Buckley considered what he called a "semirandom" distribution as the best fit to his experimental data among the six distributions studied. This is a mixed distribution, part of which is a random function and part deterministic. He did not, however, consider the log–normal distribution. Greenberg, in an

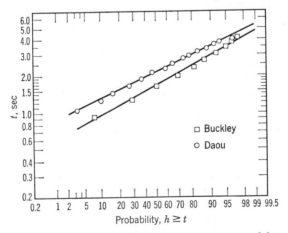

FIGURE 8. Headway distributions for two sets of data fitted by log–normal distributions.

TABLE 2

BUCKLEY DATA FITTED TO LOG–NORMAL AND SEMIRANDOM
DISTRIBUTIONS

Upper Class Interval, sec	Observed Number	Predicted Number	
		Log–Normal	Semirandom
0.90	48	62	55
1.26	132	120	114
1.62	138	129	146
1.98	112	103	111
2.34	71	72	67
2.70	35	47	41
3.06	28	29	26
3.42	16	18	17
3.78	9	11	10
4.14	4	7	7
4.14	16	11	15
Totals	609	609	609
$(chi)^2$		12.7	6.8
Degrees of freedom		8	6
Probability of worse fit		0.15	0.34

appended comment to the Daou paper,[16] made a comparison of the log–normal with Buckley's semirandom distribution, with the results shown in Table 2.

Although the semirandom shows a slightly better fit, 0.34 versus 0.15, the difference is not very significant. The semirandom distribution has four parameters while the log–normal has but two. A number of other observers have reported good fits of headway and also spacing data to a log–normal distribution.

In conclusion, one can say that the composite negative-exponential distribution and the log–normal distribution give the two best fits to experimental headway data of any distributions studied. Although not presented herein, the evidence is also available to indicate that these are the presently preferred distributions for fitting spacing data.

Breiman et al.[17] investigated the validity of a renewal hypothesis for the time headway generating process in traffic flow in one lane of a multilane highway. Statistical methods were used to determine whether time headways were independently and identically distributed. The principal conclusion of the work was that the hypothesis of independence cannot be rejected in all periods.

In his more recent work,[18] Buckley proposed a mixed distribution which he calls "semi-Poisson." One of the mixed functions is Poisson (negative exponential) and it predominates for large headways. The other function, which predominates for short headways, is selected according to those which give a good fit to small headways. The two functions found most suitable were the Gaussian and the gamma distribution in studying seven groups selected according to flow rates in increments of 5 vehicles/min, the heaviest flows being 1800 vehicles/hr and above. Group sizes ranged from a low of 694 headways to a high of 2071, a total number of more than 10,000. Although the results are interesting and the sample size large, there is little reason yet given for the advantages of such complex mixed distributions over simpler ones.

Mention should be made of the work of Underwood[19] and Miller,[20-22] who propose mixed distributions based on platooning. There is one headway distribution within platoons and an exponential or displaced exponential distribution between bunches. There is also a distribution for the number of vehicles in each bunch. This approach is discussed more fully under the heading "Moderate Traffic—Single Lane" in Section III.

Pahl and Sands[23] investigated the interaction of vehicles as a function of flow rates for each lane of a two-lane one-way highway. Their approach was to study the dependency between successive speeds caused by interaction, using speed headway operations at a point and statistical methods. Whether or not vehicles are interacting is significant in determining if they are in a car-following mode.

The method used involved a statistical comparison of the distribution of actual relative speeds with a computed distribution assuming no interaction for any of the vehicles by using random sampling from a normal distribution. Two distributions can be compared by using a modified chi-square test, adjusted to take care of the dependency between successive relative speeds such as $(v_2 - v_1)$, $(v_3 - v_2)$, $(v_4 - v_3)$, and so on; the means and variances of the distributions are not estimated.

When these necessary adjustments are made, the chi-square distribution can be used to compare actual and simulated distributions for various time headways and thereby to determine at what headways the two distributions are significantly different statistically.

The principal effect of interaction on the relative speed distribution is on the width of the distribution, which should decrease significantly as a function of the vehicles' interaction. Results showed a strong correlation between the mean value of the relative speed and the time headway up to headways of 2.5 and 4.3 sec dependent on lane and traffic flow. At low flow rates, interaction takes place at larger headways then at high flow rates. This is another illustration of the difficulty of finding general models that describe traffic behavior at all levels of flow, or concentration, and of the fact that drivers react differently in different traffic situations.

Speed Distributions

If the speeds of many vehicles are recorded as they pass a point or traverse a section of roadway, they will be found to fluctuate around some average value that depends on the concentration of traffic on the roadway. If traffic is extremely light, such that no vehicle is restrained from choosing any speed because of a slower vehicle ahead, the speeds observed are said to form a "free" speed or "desired" speed distribution. The average values and dispersions of such distributions depend on each part of the driver–vehicle–roadway complex and also on speed laws.

Some traffic flow theories seek, among other things, to explain the changes observed in speed distributions as traffic concentration increases or as vehicles travel on different kinds of roadways. In some of the circumstances studied, vehicle speeds are found to fit a normal distribution. When this is the case, the observed proportion of vehicles traveling at or less than any speed when plotted against the speed on normal probability paper will fall on a straight line. Figure 9 illustrates this result for the distribution of "free" speeds on an urban street where the mean speed is 30 miles/hr. Two other curves, idealized from data in the Highway Capacity Manual,[1] are shown to illustrate the differences between freeways and urban streets. The line for the "free" speed distribution on a freeway has a mean speed of 55 miles/hr and ranges from 1% at or less than 30 miles/hr to 99% at or less than 80 miles/hr. The urban street has a mean of 30 miles/hr and a range of 15–45 miles/hr, whereas the freeway with the same mean has a range of 22–38 miles/hr.

Until the latter part of the 1960s few experiments were made to test the range of applicability of the normal distribution to vehicle speed distributions. Also the few experiments made were not systematic studies regarding the effect of concentration, but rather partial studies regarding the location and volume of vehicles. Among the early experiments were those of Greenshields and Loutzenheiser,[24] who observed that the speed distribution is nearly normal, and those of Berry and Belmont,[25] who found that when speed is moderately limited by traffic volume, curves, or speed limits, the spot speed distribution was roughly normal. The principal differences found from the free speed distribution were lower average speed, a smaller dispersion, and a tendency for the distribution to be peaked at the center and skewed. They concluded that these deviations from normality were not serious enough to invalidate many applications of normal theory.

However, speed distributions are frequently found that do not fit a normal distribution. In a study of different types of roads in Berlin, Lindner[26] found that approximation by the normal distribution was possible only in a few

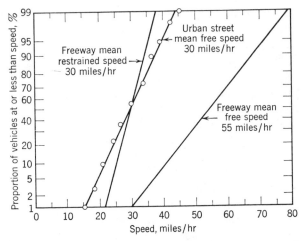

FIGURE 9. Speed Distributions on urban streets and freeways.

cases, and he suggested an expansion series with successive differentiations of the normal distribution function. Other experimenters have suggested other formulations.

However, if we assume that the mean value of a desired speed distribution, u_0, involves such parameters as roadway and ambient conditions, the distribution of speeds around u_0 will be the result of only the wishes of the drivers and the capabilities of their vehicles. With such assumptions, it is reasonable to postulate a Gaussian, or normal, distribution of free speeds as

$$f_0(v) = \frac{1}{\sigma\sqrt{2\pi}} \exp\left(\frac{-(v - u_0)^2}{2\sigma^2}\right).$$

While the speed range of the normal distribution extends from negative to plus infinity, and therefore a positive value is assigned to the probability that some vehicles will have negative speeds, for most values of u_0 and σ it is less than 10^{-6} and can be ignored. Otherwise, we have to truncate the distribution at $v = 0$, which greatly complicates the mathematics. A recent application of the Gaussian distribution was in a theoretical study of diffusion in platoons.[27]

One might list the desirable features of a generalized speed distribution as the following:

1. Must be positive.
2. Should be continuous.

3. We should find, in general, that $f_0(0)$, $f_0(\infty)$, and $f(\infty)$ all equal zero. $f(0)$ need not equal zero, since some vehicles could be forced to stop at high concentrations.
4. When the concentration approaches zero, the probability density function (p.d.f.) $f(v)$ should approach the p.d.f. $f_0(v)$, the free speed distribution.
5. Should be skewed toward positive values of v.

In addition to the Gaussian, there are other distribution functions with the desirable features. One of these is the gamma distribution, or Pearson type III:

$$f(v) = [\beta\gamma(k)]^{-1}\left(\frac{v}{\beta}\right)^{k-1}\exp\left(\frac{-v}{\beta}\right),$$

where

$$0 \leq v \leq \infty, \quad \beta > 0, \quad k > 0,$$

and

$$\gamma(k) = \int_0^\infty z^{k-1}e^{-z}\, dz.$$

The Erlang distribution is a particular case of the gamma distribution when k is an integer:

$$f(v) = [\beta(k-1)!]^{-1}\left(\frac{v}{\beta}\right)^{k-1}\exp\left(\frac{-v}{\beta}\right) \qquad 0 \leq v \leq \infty, \quad \beta > 0.$$

The Erlang distribution is encountered particularly in queuing theory.

The exponential distribution is another particular case of the gamma distribution, where $k = 1$. It has been used in traffic theory, as mentioned previously, to describe the headway distributions of a Poisson counting distribution.

The displaced gamma distribution is

$$f(v) = [\beta\gamma(k)]^{-1}\left(\frac{v-\alpha}{\beta}\right)^{k-1}\exp\left(\frac{\alpha-v}{\beta}\right),$$

where

$$\alpha \geq 0, \quad k > 0, \quad \beta > 0, \quad \alpha \geq v \geq \infty.$$

This may also be called the generalized Pearson type III. It involves all the above distributions ($\alpha = 0$) and may be used for light traffic when u_0 is at high speeds. It may also apply to a roadway when an effective enforced minimum speed exists.

The beta distribution or Pearson type I was introduced by Haight[28] as the distribution of the "constrained" speed where

$$f(v) = [B(\alpha, L\beta)L^{L\beta+\alpha-1}]^{-1}v^{\alpha-1}(L - v)^{L\beta-1}$$

$$\beta > 0, \qquad \alpha > 0, \qquad B(\alpha, L\beta) = \gamma(\alpha)\frac{\gamma(L\beta)}{\gamma(\alpha + L\beta)}.$$

The beta distribution has the interesting property of approaching a type III distribution when $L \to \infty$, that is, when the concentration tends to zero:

$$f(v) \to f_0(v) = \frac{\beta}{\gamma(\alpha)(\beta v)^{-1}} \exp(-\beta v).$$

The beta distribution could also be used where a road has both minimum and maximum enforced speed limits.

The log–normal distribution:

$$f(v) = (v\sigma\sqrt{2\pi})^{-1} \exp -\frac{(m - \log v)^2}{2\sigma^2},$$

where m and σ^2 are, respectively, the mean and the variance of $\log v$, has been used in a great variety of fields, including the analysis of economic data. Haight and Mosher[29] studied it to link the distribution of speeds over time to their distribution over space. Its very large flexibility allows it to fit a wide variety of data. For traffic flow theory it is of special interest in speed distributions as well as in headway and spacing distributions.

When one of the above does not fit a particular case, a mixed distribution may be used, as in the case of headways, where

$$f(v) = ag(v) + (1 - a)h(v)$$

and $g(v)$ and $h(v)$ are the distributions of two subpopulations. The proportion of the g type is a, and $(1 - a)$ is the proportion of the h type. This kind of distribution suggests that there are fast vehicles, described by $h(v)$, that are not constrained by slower vehicles, and slow vehicles, described by $g(v)$, which are constrained in platoons. One form of the parameter a is $a = k/k_j$, where k_j is the jam concentration occurring when vehicles are stopped. The use of mixed distributions should be avoided unless unmixed distributions are really inadequate. The existence of two subpopulations of speeds remains to be proved.

Ultimately one would hope for a theory of traffic flow that would yield realistic speed distributions. Up to this time only a few theoretical investigations have been made. Most notable and basic are those of Prigogine, Herman, and Anderson,[30] who postulate a speed distribution function $f(x, v, t)$ of a Boltzmann type derived from the principles of statistical mechanics. They have developed a kinetic theory of multilane traffic flow that has the potential of describing traffic behavior at all levels of flow:

light, moderate, and heavy. This development has taken place over a period of ten years and has been summarized in a recent monograph.[31]

While the kinetic theory employs the principles of statistical mechanics in formulating a traffic theory, one should not assume a direct application of physical Boltzmann theory. The similarity to kinetic gas theory is that both descriptions are made of the velocity distribution. The mechanisms leading to the evolution of the velocity distribution are altogether different.

Three different processes are studied in the theory. They are an interaction process, arising when a faster driver overtakes a slower one; a relaxation, or the speeding up process, expressing the attempts by drivers to reach their desired speeds; and an adjustment process, which reduces the variance around the mean speed.

These processes are expressed in a kinetic equation as

$$\frac{df}{dt} = \frac{\partial f}{\partial t} + v\frac{\partial f}{\partial x} = \frac{\partial f}{\partial t \text{ (interaction)}} + \frac{\partial f}{\partial t \text{ (relaxation)}} + \frac{\partial f}{\partial t \text{ (adjustment)}}. \quad (13)$$

This equation describes the time evolution of the speed distribution $f(v, k, t)$ of cars on a homogeneous highway at a location x at a time t when the concentration around the point is k, also a function of x, t.

The three right-hand terms are obtained by simple plausible arguments as to how the respective processes would tend to influence the speeds of the drivers involved.

The interaction process is assumed to depend on the probability of a faster car passing a slower one. This is P, which is a function of the concentration. The amount of slowing by this process is assumed to be proportional to four factors. One is $(1 - P)$, the probability of slowing down. Another is the concentration k, which means the higher the concentration, the greater the slowing effect. And the last two are the speed distribution function f giving the proportion of vehicles at a speed v, multiplied by the difference between this speed v and the mean speed of the distribution $\bar{v}$; that is, the amount of slowing depends on how fast a car is going. Thus we have

$$\frac{\partial f}{\partial t \text{ (interaction)}} = (1 - P)k(\bar{v} - v)f.$$

The relaxation process is assumed to depend on the number of vehicles that are not at their desired speed; this is $(f - f_0)$, where f_0 is the desired speed distribution. It also depends on the time lag T involved in a driver's reaching his desired speed after all constraints have been removed; T is also considered to be a function of concentration since it is reasonable to assume it would increase with concentration. Thus for the relaxation term we have

$$\frac{\partial f}{\partial t \text{ (relaxation)}} = \frac{-(f - f_0)}{T}.$$

The third term was added to the equation during development of the theory in the exploration of a singularity that was found to imply the existence of two regimes described as individual flow and collective flow, one dependent on f_0 and the other not. The collective flow regime was characterized by large fluctuations. Thus further work seeking a unified theory led to the introduction of the adjustment process

$$\frac{\partial f}{\partial t \text{ (adjustment)}} = \lambda(1 - P)k[\delta(v - \bar{v}) - f],$$

where λ is a weighting function of concentration and $\delta(v - \bar{v})$ is the Dirac delta function. This term has a simple intuitive meaning. It expresses the collective effect exerted by the local environment of a driver which tends to move his speed toward the average speed.

Although f gives the evolution of the speed distribution over time and space, the homogeneous time-independent solution is of particular interest. This solution arises when the left-hand side of Eq. 13 goes to zero, that is,

$$\frac{\partial f}{\partial x} = 0, \qquad \frac{\partial f}{\partial t} = 0, \qquad \frac{df}{dt} = 0.$$

The choice of a realistic model among the different possibilities remains essentially experimental. The procedure of gathering data should comply with certain conditions:

1. Stationarity. An important assumption of the Prigogine theory is the independence of speeds from changes in space and time so that their distribution depends only on k. In the same run of experiments, which can last a few hours, the concentration will vary, and it would be natural to partition the data by concentration groups. Data taken during rapid changes in concentration should be discarded, since the stationary state does not exist during such changes.
2. Free-flowing traffic, in addition to stationary traffic, is desirable, and the range of concentration to test is generally 0–50 vehicles/mile. For the purpose of testing the Prigogine theory, we are mostly interested in the range 0–20 vehicles/mile.
3. Homogeneity of the road. Curves, hills, weather, and so on may restrict homogeneous observations over space by photographs which, however, give a direct and convenient measure of concentration. For measurements at a point the speed distribution may also be affected by nonhomogeneous sections of the roadway upstream or downstream from the point of observation.

Locations where these conditions can be found are on straight, flat freeways, expressways, multilane rural highways, and bridges. Some freeway and expressway speed distributions are given in the *Highway Capacity Manual*,[1] pp. 49 and 50, as distribution curves. These data, which are for relatively low levels of concentration (3.6–37 vehicles/mile) are reproduced on probability coordinates in Figure 10. It is apparent from inspection that an agreement with a normal distribution is good at the range of concentrations shown.

Another set of data, in this case from the George Washington Bridge, are for relatively high levels of concentration, 35–93 vehicles/mile. These data are presented in Figure 11 in the form of probability densities. It is apparent that the normality of the distributions tends to break down above 35 vehicles/mile. Above 62 vehicles/mile the distribution tends to become bimodal. This is very pronounced at 93 vehicles/mile.

There have been very few experiments conducted to test a theory of the behavior of speed distributions. Using the data given in Figure 11, Kraus at the Port of New York Authority made such experiments, comparing the speed distributions at four concentrations with the distributions predicted by the Prigogine theory, and also with normal and log–normal distributions.[32]

The data for two concentrations, 35 and 62 vehicles/mile, are given in Tables 3 and 4, respectively. From the tables and Figure 11, one can observe a few general features of the speed distribution behavior with increasing concentration, as follows:

1. The mean speed decreases with increasing concentration.
2. The dispersion also decreases.

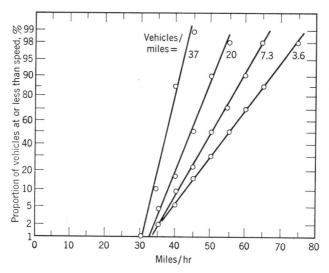

FIGURE 10. Speed Distributions on freeways and expressways at densities of 3.6, 7.3, 20, and 37 vehicles/mile.

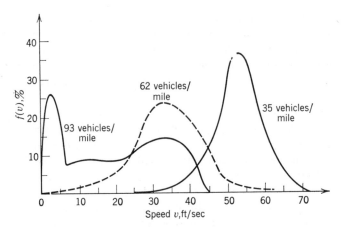

FIGURE 11. Speed distributions on the George Washington Bridge at densities of 35, 62, and 93 vehicles/mile, showing a breakdown from normality at 35 vehicles/mile into the bimodal distribution at 93 vehicles/mile as predicted by the Prigogine theory.

3. At concentrations around 60 vehicles/mile some cars are forced to zero speed.
4. Above 60 vehicles/mile the speed distribution tends to become bimodal. One part includes cars still moving; another part corresponds to local traffic jams.

All of these features are included in the Prigogine theory, but to test the theory quantitatively one must first determine what to use for the free speed $f_0(v)$ distribution. Since Figure 11 shows a tendency for the distribution to become normal at concentrations of 35 vehicles/mile or less, the $f_0(v)$ distribution was taken as normal.

Kraus made his study when the theory comprised only the interaction and relaxation process. The homogeneous time-independent solution is then given by

$$f = \frac{f_0}{1 - kT(1 - P)(\bar{v} - v)}.$$

Since T and P are function of k, the substitution $\beta = kT(1 - P)$ is made, and the speed distribution function becomes

$$f = \frac{(\sigma\sqrt{2\pi})^{-1} \exp\left[-(v - \bar{v}_0)^2/2\sigma^2\right]}{1 - \beta(\bar{v} - v)},$$

which depends on three parameters $\bar{v}_0$, σ, and β, where β is an increasing function of k. These parameters can be estimated from the data, permitting calculation of f based on the concepts of kinetic theory. Tables 3 and 4 show

TABLE 3

COMPARISON OF EXPERIMENTAL AND THEORETICAL SPEED DISTRIBUTIONS
AT $k = 35$ VEHICLES/MILE.

k	Upper Class Limit, ft/sec	Observed Number	Prigogine Theory	Normal Distribution	Log–Normal Distribution
35	<40	10	9.0	24.2	28.5
	45	25	27.2	41.3	57.0
	50	55	61.0	82.6	59.3
	55	94	83.3	79.8	83.3
	60	69	68.0	39.9	35.5
	65	26	31.0	14.3	15.1
	>65	6	5.5	2.9	6.3
	Total	285	285.0	285.0	285.0
u		52.5	52.5	52.5	—
ln (u)					3.95
σ^2		42.4			0.015
χ^2 test			$P = 0.35$	$P = 0.45$	$P < 0.005$
Kolmogorov test			$P > 0.20$	$P > 0.20$	$P > 0.20$

the results for $k = 35$ and 62. The distribution derived from Prigogine's theory fits as well as a normal distribution but no better, so that this very limited comparison with data is quite inconclusive.

More recently, other efforts have been made to test the consistency of traffic flow data with the theory developed by Prigogine and his co-workers. The first of these, by Gafarian, Munjal, and Pahl,[33] attempted to compare observed speed distributions with distributions calculated from the theory using a somewhat different approach than that just described. Their results were likewise inconclusive and they proposed modification of the theory to improve the agreement.

However, a different approach with more promising prospects involves seeing how well the theory agrees with the mean speed and the higher moments of the observed speed distributions.[34] The primary difficulty in comparing experimental results with the kinetic theory of traffic flow is that of determining the values of the parameters and the form and values of the free speed distribution. In studying only the mean speed and moments of the steady-state speed distribution the problem is that all of the parameters in the specific functions used to formulate the three processes—relaxation, interaction, and adjustment—can be consolidated into a single parameter that, for simple free speed distributions, can be related to the mean speed in a simple way.

TABLE 4

COMPARISON OF EXPERIMENTAL AND THEORETICAL SPEED DISTRIBUTIONS
AT $k = 62$ VEHICLES/MILE.

k	Class Limit ft/sec	Observed Number	Prigogine Theory	Normal Distribution	Log–Normal Distribution
62	<10	7	3.0	5.1	1.0
	15	12	10.8	18.9	4.2
	20	29	34.6	55.8	47.2
	25	69	82.6	117.0	146.0
	30	151	146.3	163.0	205.7
	35	206	190.5	215.0	180.0
	40	181	181.6	151.2	124.5
	45	129	125.1	87.5	78.2
	50	53	61.0	33.5	38.7
	55	11	19.9	10.3	18.0
	65	11	3.6	1.7	15.5
	Total	859	859.0	859.0	859.0
u		34.4			
σ^2		89.2			
χ^2 test			$P < 0.005$	$P < 0.0005$	—
Kolmogorov test			$P > 0.20$	$P > 0.20$	—

For example, if one assumes $f_0(v)$ to be a uniform distribution, where

$$f_0(v) = \frac{1}{a} \quad \text{for} \quad v_{\min} \leq v \leq v_{\max}$$
$$= 0, \quad \text{elsewhere,}$$

where

$$a = v_{\max} - v_{\min},$$

then

$$\bar{v} = \gamma + v_{\min} - a\left[\exp\left(\frac{a}{\gamma}\right) - 1\right]^{-1}.$$

The parameter γ consolidates the concentration k, probability of passing P, and the relaxation time T as follows:

$$\gamma = [(1 - P)kT]^{-1} + \lambda.$$

A simple result using these relationships is as follows:

For $\gamma = 0$, we find $\bar{v} = 25$ miles/hr $= v$ min; and as $\gamma \to \infty$, we find $\bar{v} \to 50$ miles/hr or v_0. For some intermediate value of $\gamma = 55$, the value of $\bar{v}$ is 46.3 miles/hr.

An interesting feature of the method used in this formulation is the introduction of a new variable $\eta = \gamma - \bar{v}$. One function of this variable is related to the desired speed distribution and gives a description of the driver–vehicle units on the roadway. The other, related to the parameter γ, gives a description of the interactions between driver–vehicle units in a traffic stream with a given concentration. It is from the intersection of these two functions that the mean speed is determined. Thus, although both sets of assumptions are interwoven in the theory, their effects on mean speed are separable. And it is apparent that the detailed mathematical forms of the parameters are not significant for describing multilane traffic flow in a macroscopic sense.

Furthermore, the method can be applied to a generalized model of the type proposed. The mean speed in this generalized model is again given by the intersection of two functions which are almost identical to the functions found for more specific models.[35] Numerical methods can be used with the more generalized model to find the mean speed for more complex free speed distributions.

After the mean speed has been determined, a further step can be taken by determining the other moments of the speed distribution. It is only the first few lower moments of the speed distribution that are of significance to engineering applications, the mean speed being the most important. By using the moments of the speed distribution, it is possible to formulate a set of relations between the moments that are invariant with respect to the following: the specific functional form of the desired speed distribution, the effect of the adjustment term on the variance, and the details regarding the various parameters.

Herman et al.[36] used such a method to see whether two sets of data were consistent with the theory, and concluded that conclusive results cannot be established until better and more extensive data are available. Especially needed are sufficient data to establish the speed distributions for a wide range of concentrations.

III. TRAFFIC FLOW AT DIFFERENT DENSITIES

In the Introduction, mention was made of the convenience of dividing traffic flow into three ranges—light, medium, and heavy—and qualitative descriptions were offered for each of the ranges. This breakdown is convenient because a theory which is valid in only one range of traffic intensity can give more detailed information for a particular level of complication than can a theory seeking to describe the behavior at all levels of intensity. We intuitively feel that the variable which can best be used quantitatively to define these ranges is that of the density, or concentration, of the traffic stream; that is, the amount and nature of the interactions taking place between vehicles in a traffic stream on an homogeneous roadway is primarily a function of the number of vehicles present per unit of length. We will, therefore, consider some of the theories which have been suggested for dealing with the interesting features of traffic falling within the three ranges.

Light Traffic

The simplest type of theory for light traffic derives from the assumption that the density and, consequently, the frequency of interactions between vehicles are essentially zero. A theory of this kind, in which a driver maintains his own desired speed at all times, shows that any distribution of vehicles along a highway will tend toward the Poisson distribution.

Weiss and Herman[37] showed this in a study that considered an infinitely long line of traffic moving on a homogeneous highway without traffic lights. It is assumed that each vehicle travels at a constant speed that is a random variable, and that when one vehicle overtakes another, passing is always possible without a change in speed. This is consistent with light traffic flow.

The model consists of a single infinitely long lane of homogeneous traffic with cars of zero length. It is assumed that at $t = 0$, cars are set down on the highway so that the probability of a car being located at a distance x from a given marked car is $p(x, 0)$ and that

$$p(x, 0) = p(-x, 0).$$

The expression for $p(x, t)$ is given by

$$p(x, t) = \int_0^\infty \int_0^\infty p(x + (v - v')t, 0)n(v)n(v') \, dv \, dv'$$

as

$$t \to \infty, \qquad p(x, t) \to p \text{ (a constant)},$$

where $n(v)$ is the probability density function of speed v of the marked car and $n(v')$ that of a second car.

If $n(v)$ is chosen to be a combination of two delta functions

$$n(v) = a\delta(v - V) + (1 - a)\,\delta(v - V^*),$$

then

$$p(x, t) = [a^2 + (1 - a)^2]p(x, 0) + a(1 - a)p(x + (V - V^*)t, 0)$$
$$+ a(1 - a)p(x + (V^* - V)t, 0).$$

In a slightly more complicated theory of light traffic behavior, one gives effect to the fact that when the concentration is greater than zero there will be occasions when a faster driver overtakes a slower one and is delayed in passing. If density is sufficiently low but still not zero, one can ignore the effects of interactions involving more than the two vehicles. Theories of this type were suggested by Newell,[38] Bartlett,[39] and Carleson.[40] All of these theories lead to the same conclusion, namely, that the time average velocity v of a car with a desired speed v_f is less than v by an amount proportional to the concentration k, that is,

$$v = v_f(1 - ak),$$

where a, the constant of proportionality, is sometimes taken to be the jam spacing or reciprocal of the jam concentration, namely, $a = s_j = 1/k_j$. Greenshields, who found this linear relationship in empirical studies, believed it would hold right up to the jam concentration. More recent studies and theoretical considerations show that it holds only for light traffic on certain roads. The Greenshields data were taken primarily on two-lane rural highways, one type of roadway where this theory is valid.

On multilane freeways experiments reveal a quite different behavior: The mean speed is insensitive to increasing concentrations until they reach medium or heavy traffic levels.[40,41] Newell,[92] therefore, assumed that on such roadways, the average speed at moderate densities is due to interactions involving three or more vehicles simultaneously. If a highway has only two unidirectional lanes, then both lanes are temporarily blocked when a second vehicle is passing another. If a third vehicle overtakes the pair, it will be delayed until the passing lane becomes available. For a three-lane freeway, three vehicles could block the passing of a fourth vehicle.

Newell studied the behavior of car 0, which has a desired speed v_0, which it maintains except during a passing involving three cars. He derived expressions for the average speed of car 0, $u(v_0)$ and of all cars, $u(v)$ as follows:

$$u(v_0) = v_0 - k^2(D_1 + D_2)^2 \sigma G\left(\frac{v_0 - V}{\sigma}, \beta\right),$$

$$u(v) = V - k^2(D_1 + D_2)^2 \sigma G(\beta),$$

where

$V = E(v_0)$, mean free speed,

k = concentration,

D_1 = distance behind a car ahead when a passing maneuver starts,

D_2 = distance ahead of car passed when passing maneuver is completed,

G = a nondimensional function.

The parameter β gives the speed assumed by a car 1, which is passing a car 2 when it is overtaken by car 0, as follows:

$$v_1^* = (1 - \beta)v_0 + \beta v_1.$$

In contrast with models that deal only with passings involving two cars and predict a linear decrease of u with k, Newell's model, which includes three-car interactions, predicts that u decreases proportional to k^2.

Moderate Traffic—Single Lane

Several ways for dealing with moderate traffic have already been discussed in connection with speed and headway distributions. These, however, with the exception of the Prigogine theory, have little theoretical content. The Prigogine theory is not limited to moderate traffic, but instead seeks to show the speed distribution for all traffic intensities. Theories seeking to deal specifically with moderate traffic have been developed using two basically different approaches. One is to extend the analogy with the kinetic behavior of gases, which proved useful for light traffic. The other is to apply queuing theory.

In extending the kinetic gas analogy to moderate traffic flow, several simplifying assumptions made with light traffic are no longer acceptable. For example, one cannot assume that the probabilities of various kinds of interactions between various numbers of cars can be neglected. A first-order theory for light traffic can justify the use of a Poisson distribution in order to estimate passing rates and the clustering of vehicles, but this is no longer valid for moderate traffic. The delays to vehicles result in a higher probability of finding two or more cars together, particularly fast cars behind slow ones because the interactions cause cars to stay together for a while. The necessary use of distributions for pairs or larger groups of cars adds complications, and these interactions give rise to correlations between velocities and positions. For example, if three cars have interacted simultaneously and the two fastest cars pass the slowest one, a complication arises in that the expected rate of passing in the future is not independent of the past. If the third car has a desired speed greater than the intermediate car, which passes first, it is obvious that the faster car will soon want to pass the intermediate car after both have passed the slowest one.

Instead of trying to include all the various interactions possible for a group of vehicles, some theorists, notably Tanner[42] and Komentani[43] have simplified the problem by studying the behavior of just one vehicle in a system of vehicles, the behavior of all the other vehicles being assumed to be already known. Other authors, Newell,[38] Carleson,[40] Bartlett,[39] Prigogine,[30] and Miller[20] have attempted to consider all of the vehicles as forming one large interacting system. The first approach, being easier to handle, allows for detailed and realistic models, while the second yields sets of simultaneous differential equations compelling a resort to models that are fairly primitive. Carleson and Bartlett assumed random passing with no group of vehicles larger than two. A most fruitful approach to a theory of moderate traffic is that of Miller, which we shall describe. It deals with two-lane rural roads.

Miller's approach to moderate traffic flow is that of queuing theory. An early paper[21] gives the essentials of his approach and has the virtue of relative simplicity compared to later work, which is somewhat more realistic. Miller's early model incorporates the following assumptions:

1. Vehicles travel in clusters or queues. A queue may consist of one or more vehicles. The queues are independent of each other in size, position, and velocity.
2. Queues of one vehicle are represented as points, without physical length. A correction can subsequently be made to change this assumption.
3. The rate of catching up of queues is proportional to the product of the concentrations of the queues and their velocities.
4. Overtaking (the successful passing of a queue leader by a follower within the queue) occurs at a rate which is a function of queue size and velocity.
5. Traffic in one direction is assumed to be stochastically independent of traffic in the other direction.
6. The identity of an individual vehicle is not preserved through time. Thus, for example, an overtaking vehicle will freely select its speed from the distribution $h^*(u, v)$, defined below, independently from the value of its free speed at any earlier time.
7. The speed of a vehicle changes instantaneously when it catches up and when it overtakes a queue.

DEFINITIONS.

$k_i(t)$ = density (number per unit road length) of queues of i vehicles.

$h_i(u; t) \, du$ = probability that the velocity of an i-vehicle queue is in the range $(u, u + du)$ at time t.

$\lambda_i(u; t) \, dt$ = probability that a vehicle overtakes and leaves an i-vehicle queue in the time interval $(t, t + dt)$, given that the queue velocity is u.

$h^*(u, v; t) \, du$ = probability that a vehicle, overtaking from a queue with velocity v at time t, then takes a velocity in the range $(u, u + du)$.

THE CASE OF $i > 1$.   We have

$$k_i(t)h_i(u; t)\, du = \text{density of } i\text{-vehicle queues with velocity}$$
$$\text{in } (u, u + du).$$

In a short time interval dt,

$$k_i(t + dt)h_i(u; t + dt)\, du - k_i(t)h_i(u; t)\, du \tag{14}$$

is the change in dt in density of i-vehicle queues with velocity in $(u, u + du)$. There are four contributions to this change (as usual, we ignore terms of second or higher order):

1. Change due to vehicles overtaking from queues of i vehicles:

$$[-\lambda_i(u; t)\, dt] \cdot [k_i(t)h_i(u; t)\, du].$$

2. Change due to vehicles overtaking from queues of $(i + 1)$ vehicles:

$$[+\lambda_{i+1}(u; t)\, dt] \cdot [k_{i+1}(t)h_i(u; t)\, du].$$

3. Change due to r-vehicle queues catching up to $(i - r)$ vehicle queues. Consider lead queues with velocity in the range $(u, u + du)$ and follower queues with velocity in the range $(u', u' + du')$. A follower may catch up if $u' > u$. For a follower queue to catch up with a lead queue in dt, the distance between the two queues must be less than $(u' - u)\, dt$ at the beginning time t. The proportion of total roadway with the property that such follower queues will catch up in dt is

$$
\begin{bmatrix}
\text{Number of lead queues, per} \\
\text{unit length of road, with} \\
(i - r) \text{ cars and velocity in} \\
\text{the range } (u, u + du).
\end{bmatrix}
\times
\begin{bmatrix}
\text{roadway length, per} \\
\text{lead queue, on which a} \\
\text{follower queue must be} \\
\text{at time } t \text{ in order to} \\
\text{catch up in } dt
\end{bmatrix}
$$

$$= [k_{i-r}(t)h_{i-r}(u; t)dt] \times [(u' - u)\, dt].$$

The overall density of follower queues of length r and, velocity in the range $(u', u' + du')$ is

$$k_r(t)h_r(u'; t)\, du'.$$

These followers are assumed to be distributed uniformly along the road. Therefore:

The increase in dt of density of i-vehicle queues, due to r-vehicle queues, with velocities in the range $(u', u' + du')$, catching up to $(i - r)$-vehicle queues with velocities in the range $(u, u + du)$

= the change in dt of density of r-vehicle follower queues, with velocities in the range $(u', u' + du')$, due to catching up with $(i - r)$-vehicle queues with velocities in the range $(u, u + du)$

= [overall density of such follower queues] · [fraction of roadway with the property that the catching up can occur]

= $k_{i-r}(t)h_{i-r}u; (t) du (u' - u) dt k_r(t)h_r(u'; t) du'$.

To obtain the total increase in dt, of density of i-vehicle queues, by such catchings up, we sum over all possible values of r and integrate over all $u' > u$:

$$\left[\sum_{r=1}^{i-1} k_{i-r}(t)h_{i-r}(u; t)k_r(t) \int_u^\infty h_r(u'; t)(u' - u) \, du' \right] du \, dt$$

4. Change due to i-vehicle queues catching up or being caught up by other queues. The derivation is similar to that for 3 and yields

$$\left[k_i(t)h_i(u; t) \sum_{r=1}^\infty k_r(t) \int_0^\infty h_r(u'; t) |u' - u| \, du' \right] du \, dt.$$

The total change in dt, of density of i-vehicle queues in the range $(u, u + du)$ is the sum of these four terms. This sum is equated to Eq. 14. Taking limits as $dt \to 0$, the equation reduces to the following general equation for $i > 1$:

$$\frac{d}{dt} [k_i h_i(u)] = \sum_{r=1}^{i-1} k_{i-r}h_{i-r}(u) \int_u^\infty k_r h_r(u')(u' - u) \, du'$$

$$+ \lambda_{i+1}k_{i+1}h_{i+1}(u) - \lambda_i k_i h_i(u) - k_i h_i(u)$$

$$\times \sum_{r=1}^\infty \int_0^\infty k_r h_r(u') |u' - u| \, du'. \tag{15}$$

THE CASE $i = 1$.   The density of queues of one vehicle may increase or decrease in three ways:

1. By a vehicle overtaking from a queue of two vehicles;
2. By queues of one vehicle catching up or being caught by other queues;
3. By the one-vehicle queues created by the overtaking vehicles from any size queues.

The derivation of the general equation is similar to that for the case $i \neq 1$. The result is

$$\frac{d}{dt} [k_1 h_1(u)] = \lambda_2 k_2 h_2(u) - k_1 h_1(u)$$

$$\times \sum_{r=1}^\infty \int_0^\infty k_r h_r(u') |u' - u| \, du'$$

$$+ \sum_{r=2}^\infty k_r \int_0^u \lambda_r h^*(u, v)h_r(v) \, dv. \tag{16}$$

It is evident that the quantitative solution to Eqs. 15 and 16 requires substantial experimental information regarding the passing mechanism and the speed distributions of both isolated and queue-leader vehicles.

However, some approximate results can be derived from simple assumptions. If, for example, we assume that the rate of overtaking queues is proportional to the product of their concentration and their relative velocities, we can derive the following:

$$\frac{dk}{dt} = -ak^2 + \lambda(k - k_1),$$

where k is the concentration of all queues plus single vehicles, λ is the rate at which vehicles leave queues, and the constant a turns out to be one-half the mean difference in velocity. For a stream in equilibrium ($dk/dt = 0$), the proportion of single vehicles is

$$\frac{k_1}{k} = 1 - \frac{ak}{\lambda},$$

and the term ak/λ is a measure of bunching.

It should be noted that any lack of realism resulting from assumption 2 (vehicles treated as points) can be corrected easily. Miller points out that the relationship between concentration in his model and the real world can be derived as follows. Suppose that in practice the average distance between consecutive queuing vehicles is b. Then if in the real traffic there are N vehicles on a length of road X, in the corresponding model which preserves distances between queues of vehicles, there must be N vehicles on a length of road $X - Nb$. If the concentrations of vehicles are k and k' in the model and real traffic, respectively, we have

$$k = \frac{N}{X - Nb} = \frac{N/X}{1 - Nb/X} = \frac{k'}{1 - k'b},$$

and conversely $k' = k/(1 + kb)$. A very similar relationship can be derived for rates of flow in the two systems.

Moderate Traffic-Multilane

When there are two or more lanes in one direction, traffic theorists have been more interested in the interactions taking place between the lanes rather than the passing and clustering phenomena. We will look at two papers dealing with this problem.

The first paper by Gazis, Herman, and Weiss appeared in 1962.[44] It deals with the density oscillations between parallel lanes and investigates this phenomenon by means of a simple mathematical model. Among the aspects investigated is that of stability, that is, the attenuation of disturbances from an equilibrium density distribution. A series solution is obtained for a system of differential difference equations with a time lag. The solution is applicable to other problems described by similar equations. It is derived for two lanes and is then generalized to n lanes. As $n \to \infty$, the instability doubles for that of two lanes.

The assumptions of this approach are

1. Conservation of vehicles,
2. Existence of an equilibrium density,
3. Independence of density with location,
4. Rate of exchange of vehicles between adjacent lanes is proportional to the difference in the number of vehicles in the two lanes minus the difference for the equilibrium state.
5. Rate of exchange lags this difference by a constant time lag T.

These assumptions lead to the following mathematical statement of the theory:

$$\dot{k}_1(t) = -\dot{k}_2(t) = \lambda[k_2(t - T) - k_1(t - T) - (k_{20} - k_{10})], \qquad (17)$$

where

$$\dot{k}_i(t) = \text{time rate of change of density in lane } i \text{ at time } t,$$

$$k_i(t - T) = \text{density in lane } i \text{ at time } (t - T),$$

$$k_{io} = \text{equilibrium density in lane } i \ (i = 1, 2),$$

$$\lambda = \text{constant of proportionality.}$$

The two equations can be reduced to the differential difference equation:

$$\dot{\sigma}(t + T) = -2\lambda\sigma(t), \qquad (18)$$

where $\sigma = $ the deviation in lane 1 from its equilibrium value and $-\sigma$ the deviation in lane 2, that is,

$$\sigma(t) = k_{10} - k_1(t),$$
$$-\sigma(t) = k_{20} - k_2(t).$$

The stability of $\sigma(t)$ can be investigated by Laplace transforms with the following results:

1. If $2\lambda T \leq e^{-1}$, $\sigma(t)$ is damped, nonoscillatory, and decreases exponentially with t.
2. If $e^{-1} < 2\lambda T < \pi/2$, $\sigma(t)$ is damped and oscillatory.
3. If $2\lambda T = \pi/2$, $\sigma(t)$ is oscillatory with constant amplitude.

4. If $2\lambda T > \pi/2$, the amplitude of the oscillations increases with time and $\sigma(t)$ is unstable.

The stability of n lanes is given by multiplying $2\lambda T$ by the expression $(1 - \cos \pi/n)$ and as $n \to \infty$ the point of instability is at $4\lambda T > \pi/2$.

The question of the values of the equilibrium states k_{10} and k_{20} (and a number of other questions) were investigated by Oliver.[93] He investigated the continuity conditions where the following hold

$$\frac{\partial k_1}{\partial t} + \frac{\partial q_1}{\partial x} = p_{21}(x, t) - p_{12}(x, t),$$

$$\frac{\partial k_2}{\partial t} + \frac{\partial q_2}{\partial x} = p_{12}(x, t) - p_{21}(x, t).$$

These expressions state that the net effect of time changes in density and space changes in flow at a point x and a time t are equal to the exchange between the two lanes in a differential section of roadway Δx. If there is no exchange, the equations become the well-known one-dimensional continuity equation

$$\frac{\partial k}{\partial t} + \frac{\partial q}{\partial x} = 0.$$

Oliver studied a model in which lane changes were proportional to the square of the density in the lane that was the source of departing vehicles and the first power of the emptiness (unused space) of the lane-admitting vehicles. Thus

$$P_{12}(x, t) = ak_1^2(x, t)[k_{2j} - k_2(x, t)], \tag{19}$$

$$P_{21}(x, t) = bk_2^2(x, t)[k_{1j} - k_1(x, t)], \tag{20}$$

and k_{1j} and k_{2j} are the jam (or maximum allowable) densities in the respective lanes.

The choice of this lane-changing function was partly theoretical but largely empirical.

Equations 19 and 20 can be solved for equilibrium conditions where density and flow may be position-dependent but not time-dependent. The solution is complex and will not be given here.

Holland[45] developed a computer model for multilane flow which would determine a variety of steady-state quantities for four-lane and six-lane highways with a center dividing strip between opposing traffic flows.

With a given set of free speeds $v_1, v_2, \ldots, v_n$, where n is any finite number, a class i car travels at speed v_i until it encounters a queue of one or more slower cars. There are ρ_j queues per unit distance traveling at speed v_j in the right-hand lane, and e_j cars per unit distance and unit time that enter the roadway and leave the roadway in a steady state.

From these parameters Holland determines:

1. Composition of queues in each lane (n_{ij} = the mean number of i cars in a j queue) as a function of the speed class of queue leaders,
2. The queue densities (ρ_{ij}) for all lanes except the right lane.

From these basic quantities, one can determine most of the steady-state quantities of interest.

The basic assumptions of the model are

1. The arrival of queues behind slower queues is a Poisson process.
2. Cars enter and leave in a Poisson manner.
3. Cars travel at their free speed between queues.
4. There is a relaxation process between queues.
5. Only cars impeded by slower cars pass into the adjacent lane to their left (slower vehicles tend to the right).
6. The time spent in a queue is exponentially distributed; the rate of passing depends on queue speed, desired speed, and the characteristics of the adjacent lane.

Despite the limitations of certain approximations, limited comparisons with empirical data indicated that the model provided reasonable predictions of the steady-state behavior on multilane highways.

Heavy Traffic

A great deal of traffic flow theory is concerned with heavy traffic flow for which there have been the two basic approaches mentioned in the Introduction to this chapter. These, the microscopic point of view of car-following theory and the macroscopic point of view of kinematic wave theory, are of such importance that a section of this chapter is devoted to each. We first consider the latter.

IV. TRAFFIC DYNAMICS—MACROSCOPIC

Macroscopic theories of traffic flow treat the traffic stream as a continuous fluid. The behavior of individual vehicles is ignored and one is concerned only with the behavior of sizable aggregates of vehicles. Vehicular traffic is not very much like the physical fluids we know; yet there are sufficient similarities to make hydrodynamic theory useful in describing traffic dynamics. As a fluid, traffic has particles weighing tons, or millions of grams, while an average molecule of air is weighed in micro-micro-micro-micro grams. The traffic particles are not only extremely heavy but also are relatively inelastic compared to molecules, which are perfectly elastic. Fortunately, vehicles are relatively ingenious in avoiding the frequent collisions that occur in physical fluids and that account for much of their macroscopic behavior. Traffic streams are extremely concentrated compared to ordinary gases. In 1200 cm³ of air, the amount of space occupied by molecules is only 2 cm³. Thus the "occupancy" is only $2/1200 = 0.00167$, compared with a nominal traffic "occupancy" of 0.16, 100 times as concentrated.

With differences of such magnitudes, it is surprising to find that kinetic theories of gases would be applicable to traffic streams. This shows the power and applicability of scientific laws like the conservation of matter and energy. If we assume that vehicles are neither created nor destroyed in traveling down a roadway, then the law of conservation will apply. We know that cars are not created on the roadway, and their destruction is, fortunately, rare, so in general, matter and energy are conserved. This law alone leads to an understanding of many of the qualitative features of traffic dynamics, especially in the formation and propagation of waves.

Hydrodynamic Theory

The first and most important contribution to the macroscopic theory of traffic flow was made in the classical paper by Lighthill and Whitham,[46] who drew an analogy between the flow of traffic and the behavior of flood movements in rivers and supersonic air flow around projectiles.

The fundamental hypothesis of the theory is that at any point of the road the flow of (vehicles/hour) is a function of the concentration k (vehicles/mile). The hypothesis implies that a slight change in flow is propagated back through the stream of vehicles along a kinematic wave whose velocity relative to the road is the slope of the graph of flow versus concentration. It has been shown for fluids that kinematic waves can run together and form kinematic shock-waves at which fairly large reductions in speed occur very quickly. These too are common on roadways, notably at the rear of a slow moving platoon or a queue of vehicles behind a bottleneck.

The flow q and the concentration k have no significance except as means; the purpose of the theory is to determine how these mean values vary in space and time. We seek to make this determination by considering the speed with which changes in q and k are propagated along the roadway.

A fixed observer sees a flow $q = N/T = uk$, where N is the number of vehicles passing him in a time period of duration T, and u, of course, is the mean speed (space or harmonic) at which the N vehicles pass him. If the observer is not stationary but instead moves upstream at a uniform speed c, he will pass an additional number of vehicles ck in addition to q so that

$$q + ck = \frac{N}{T}.$$

If he moves down stream this becomes

$$q - ck = \frac{N}{T},$$

and N is the difference between the number of vehicles passing the observer and those which he passes. If he moves at the mean speed of the stream, then $N = 0$ and $c = u$.

Consider now two observers moving with a uniform speed c, the second starting and remaining a time T behind the first. Suppose that the flow and concentration are changing with time, but that nevertheless the observers jointly adjust their speed c so that the number of vehicles which pass them minus the number which they pass is, on the average, the same for each during a time interval T. This is illustrated in Figure 12. The result of their observations would be

$$q_1 - ck_1 = \frac{N}{T} = q_2 - ck_2,$$

if they were moving downstream in the positive direction of flow. Solving these for c, we find that

$$c = \frac{q_2 - q_1}{k_2 - k_1}.$$

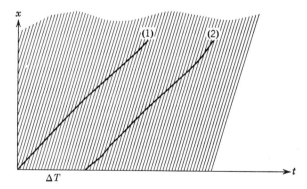

FIGURE 12. Vehicles passed by two moving observers.

If the changes in flow and concentration are small, then

$$c = \frac{\Delta q}{\Delta k},$$

which in the limit becomes

$$c = \frac{dq}{dk}.$$

Thus when the difference in flow and concentration are small, they are propagated at a speed given by the tangent dq/dk to the curve q versus k.

There is a more easily understood way of deriving the wave celerity. This is illustrated in Figure 13, where a concentration k_1 is moving down the roadway $0 - x$ behind lesser concentration k_2 at a speed c. At the boundary S between the two concentrations, where there is a shockwave, we assume a conservation of vehicles, that is, the number entering the moving shockwave S must equal the number exiting; the flow in must equal the flow out. Assuming that the wave speed c is less than the space mean traffic speed on each side, we have

$$\text{flow in} = (u_1 - c)k_1$$

$$\text{flow out} = (u_2 - c)k_2.$$

Solving for c, $c = (q_2 - q_1)/(k_2 - k_1)$ and for small disturbances, $c = dq/dk$.

This relationship is particularly useful when $q = q(k)$ at a point on the roadway, that is, when q is a function of k only at any point. If this function is known, then the speed at which a small disturbance is propagated along the roadway is known if either k or q is known. The flow-concentration curve is one of the most useful relationships to study because the slopes of the tangents and the chords on the q–k curve show the speed of wave propagations, while the slopes of the radii vectors give stream speeds. Such slopes may be directly translated to a space–time x–t diagram, as illustrated in Figure 14. By a proper choice of coordinate scales, the same speed will have the same slope on both graphs. This is especially useful in dealing with

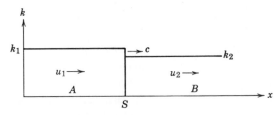

FIGURE 13. Celerity c of a wave carrying an increase in concentration down a roadway.

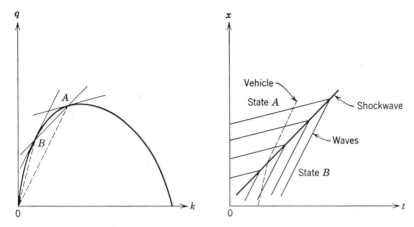

FIGURE   14. Relationship   between   flow–concentration   and   space–time coordinates.

problems graphically without having to solve complicated differential equations of motion in order to deduce what traffic events will occur in space and time for any boundary conditions.

Figure 14 merits some discussion. The $q–k$ curve describes various steady states of a traffic stream at some point x on the roadway. If a section of a roadway is homogeneous, a particular $q–k$ curve will apply over the entire homogeneous section. Figure 14 assumes that the roadway $0–x$ is homogeneous. The traffic stream is changing from the state represented by the point A on the $q–k$ curve to the state represented by the point B. This could happen because of vehicles in light, high-speed traffic catching up with slower more concentrated vehicles ahead. The waves carrying the higher speed flow in state B overtake the slower waves carrying the state A down the roadway. When the two groups of waves run together, they form the shockwave whose trajectory is given by the heavy line in the $x–t$ diagram. The speed of the shockwave is given by the slope of the chord $B–A$, joining the two states on the $q–k$ curve. Ahead of the shockwave the waves travel at a speed given by the slope of the tangent to the $q–k$ curve at point A. Behind the shockwave, the waves travel at a much faster speed represented by the slope of the tangent to the $q–k$ curve at point B.

The trajectories of the waves must not be confused with the trajectories of the vehicles in the traffic stream. Since the mean speed of a vehicle in a traffic stream of state q, k is given by the relationship $u = q/k$, the vehicle speed is represented geometrically by the slope of the radius vector to the state point. The speed of a vehicle in state A is given by the slope of the line $0–A$ in the $q–k$ coordinates; the speed of a vehicle in state B, by the slope of the line

0–*B*. The trajectory of the average vehicle may be shown on the space–time diagram for state *B*, by lines parallel to 0–*B*; and for state *A*, by lines parallel to 0–*A*. Such a vehicle passing from state *B* to state *A*, thereby going through the shockwave, is shown by the dashed line on the *x–t* diagram of Figure 14.

Slightly more complicated changes of state are portrayed in Figure 15. Here there is a gradual increase in the in-flow to the roadway at point 0, followed by a gradual decrease back to the initial level. The rise and fall of the in-flow can be measured by an observer at point *x* = 0 who counts the number of vehicles passing in equal intervals of time. The question is, can one predict what observations will be made along the roadway as the surge passes down it? When will it reach a given point? Will it spread out or become more compressed in time and space? What will be the effect on speeds and travel times?

The wave theory of Lighthill and Whitham gives the answers to these questions in the simple graphical constructions of Figure 15. The wave paths at any time are parallel to the tangent to the left side of the *q–k* curve at the point on the curve having a value of *q* equal to that counted by the observer at *x* = 0. As the in-flow increases, the waves travel more and more slowly, thus tending to separate and fan out as they pass down the roadway. However, when the flow decreases successive waves travel faster and faster and therefore compress and run together.

A shockwave starts to form at the point where two waves first run together. Subsequently its trajectory can be traced by making it parallel to a chord on the *q–k* curve joining the two states represented by intersecting flow waves. Since the slope of this chord is approximately the mean slope of the inter-secting flow waves, the shockwave path can be reasonably well approximated by estimating the mean slope by eye without transferring it from the *q–k* curve. This is the method used by Lighthill and Whitham to construct Figure 15.

It can be seen that the traffic surge up to the point *DE* has a constant direction but then diffuses as it passes down the roadway. The front of the surge moves with the speed of the initial waves, while the rear of the surge starts at the speed of the slowest waves but increases in speed as the fan of faster waves overtake it.

It should be iterated that lines drawn in Figure 15 are waves, that is, lines of constant flow. For a uniform road they are lines of constant speed. The average speed of the traffic stream is faster than the wave speed. Most vehicles will travel faster than the waves and therefore travel through the heavier flow. The dotted line in Figure 15 shows the trajectory of a car which travels at the average traffic speed. On entering the traffic hump a driver has to slow down fairly rapidly as he approaches the shockwave. After leaving it he increases his speed slowly as he traverses the fan of waves downstream.

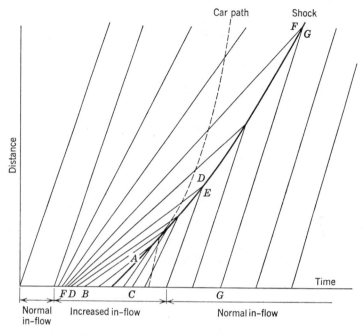

FIGURE 15. Behavior of a traffic surge.

Richards[47] studied the propagation of waves using the hydrodynamic analogy independently of Lighthill and Whitham. He likewise based his analysis on the classical continuity equation

$$\frac{\partial k}{\partial t} + \frac{\partial q}{\partial x} = 0.$$

Figure 16 illustrates on a space–time diagram how the conservation of vehicles on a section of roadway Δx during a time interval Δt results in the above equation. Vehicles enter the rectangular space–time domain Δx–Δt, from the lower and left sides and leave through the upper and right sides.

In addition to the continuity equation, Richards assumed a particular equation of state at each point on a homogeneous roadway. The equation of state $q = q(k)$ resulted from an assumption that the space mean traffic speed is a linear function of concentration, namely,

$$u = u_f\left(1 - \frac{k}{k_j}\right).$$

(See discussion of this relationship in the part of Section III on light traffic.) Richards further assumed the general relationship $q = uk$.

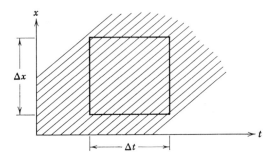

FIGURE 16. Illustration of continuity of cars;
$q\Delta t + k\Delta x = (q + \Delta q)\Delta t + (k + \Delta k)\Delta x$,
hence, $(\Delta q/\Delta x) + (\Delta k/\Delta t) = 0$.

With these assumptions Richards derived the formula for k as a function of x and t and the parameters u_f, the mean free speed, and k_j the jam concentration. The derivation is as follows:

$$q = uk = u_f\left(k - \frac{k^2}{k_j}\right),$$

$$c = \frac{dq}{dk} = u_f\left(1 - \frac{2k}{k_j}\right),$$

$$\frac{\partial q}{\partial x} = u_f\left(1 - \frac{2k}{k_j}\right)\frac{\partial k}{\partial x}, \qquad (21)$$

$$\frac{\partial k}{\partial t} + u_f\left(1 - \frac{2k}{k_j}\right)\frac{\partial k}{\partial x} = 0.$$

The solution of Eq. 21 is an arbitrary function of the argument, that is,

$$k = f\left[x + u_f\left(\frac{2k}{k_j} - 1\right)t\right].$$

While Richards did not solve this equation explicitly, he pointed out that for $t = 0$, $f(x) = k(x, 0)$, and for a fixed value of $k = k_0$, the argument of the arbitrary function is a constant and k_0 propagates at a constant speed. Therefore

$$x = x_0 - u_f\left(\frac{2k_0}{k_j} - 1\right)t = x_0 - ct$$

gives the location of the concentration k_0 at a time t when its location at $t = 0$ was x_0. Two different cases are presented in Figures 17 and 18.

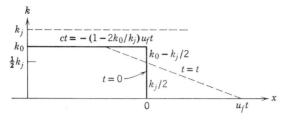

FIGURE 17. Concentration versus space for traffic accelerating to the free speed u_f from a concentration of k_0.

Figure 17 shows a vehicle pulling away at $x = x_0 = 0$ from a group of vehicles with a concentration downstream and the decreasing concentration upstream from $x = x_0$ at any arbitrary time t after the first vehicle pulls away from the group at the mean free speed u_f. The wave speed is $c = u_f(1 - 2k_0/k_j)$. Because of the geometry, the line of changing concentration always passes through the point x_0 and $k_j/2$.

Figure 18 illustrates wave propagations caused by a traffic stoppage of duration T with an initial concentration of k_0. The wave velocity is $c = \Delta q/\Delta k = -u_f(k_0 - k_0^2/k_j)/(k_j - k_0) = -u_f k_0/k_j$. The starting wave moves at a speed $-u_f$, thus the relative speed of the two waves is $u_f[(1 - k_0)/k_j]$, and the initial distance between them is $u_f(k_0/k_j)T$. The time the starting wave overtakes the stopping wave is then

$$t_0 = \frac{u_f(k_0/k_j)T}{u_f(1 - k_0/k_j)},$$

$$t_0 = \frac{T}{(k_j/k_0 - 1)}.$$

The solid line in Figure 18 shows the distribution of vehicles on the roadway at $t = 0$. The roadway has already been blocked at $x = 0$ for a time T and the jam concentration k_j extends upstream to a point

$$x = -cT = -\frac{k_0}{k_j} u_f T.$$

At $t = 0$, we assume that the initial stoppage is released and a starting wave moves upstream at a speed u_f, its location being given by $x = u_f t$. The dashed lines show the distribution of concentration along the roadway at two times: t_1, before the starting wave overtook the stopping wave; and t_2, after the overtaking. It will be noted that the concentration at $x = 0$ remains at the level $k_j/2$ and flow stays at the maximum level of $u_f k_j/4$ as the starting and other waves move upstream.

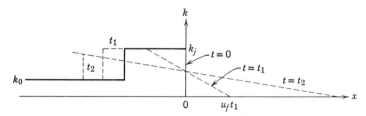

FIGURE 18. Concentration versus space for traffic behind and in front of a traffic jam. Stoppage released at $t = 0$, wave travel at times t_1 and t_2.

The progress of the vertical face, or barb, of the arrowhead-shaped concentration profile, once the starting wave overtakes the stopping wave, is determined by the wave speed

$$c = -\frac{dx}{dt} = \frac{\Delta q}{\Delta k} = \frac{u_f(k - k^2/k_j) - u_f(k_0 - k_0^2/k_j)}{k - k_0},$$

$$\frac{dx}{dt} = -u_f\left(1 - \frac{k}{k_j} - \frac{k_0}{k_j}\right). \tag{22}$$

By similar triangles

$$\frac{k - k_j/2}{x} = \frac{(k_j/2)}{u_f t},$$

$$\frac{k}{k_j} = \frac{x}{2u_f t} + \frac{1}{2}.$$

Substituting in Eq. 22, we find

$$\frac{dx}{dt} = -u_f\left(\frac{1}{2} - \frac{x/2}{u_f t} - \frac{k_0}{k_j}\right),$$

$$\frac{dx}{dt} - \frac{x}{2t} = u_f\left(\frac{k_0}{k_j} - \frac{1}{2}\right),$$

$$xt^{-1/2} = 2u_f\left(\frac{k_0}{k_j} - \frac{1}{2}\right)t^{1/2} + c_1,$$

$$x = 2u_f\left(\frac{k_0}{k_j} - \frac{1}{2}\right)t + c_1 t^{1/2}.$$

The integration constant c_1 can be evaluated from the conditions when $t = t_0$, $x = -u_f t_0$,

$$c_1 = 2u_f\left(T\frac{k_0}{k_j}\left(1 - \frac{k_0}{k_j}\right)\right)^{1/2},$$

and

$$x = 2u_f \left\{ \left(\frac{k_0}{k_j} - \frac{1}{2} \right) t + \left[T \frac{k_0}{k_j} \left(1 - \frac{k_0}{k_j} \right) t \right]^{1/2} \right\}. \tag{23}$$

The peak concentration is

$$k_{max} = k_0 + \left(T \frac{k_0}{k_j} \frac{(1 - k_0/k_j)}{t} \right)^{1/2}$$

and eventually $k_{max} \rightarrow k_0$ as $t \rightarrow \infty$.

If $k_0/k_j < \frac{1}{2}$, the t term in Eq. 23 will eventually dominate the $(t)^{1/2}$ term. Therefore the shockwave will eventually reverse its motion and return to the origin for the condition $x = 0$ in Eq. 23. The time of return is

$$t_r = T \frac{k_0}{k_j} \frac{(1 - k_0/k_j)}{(\frac{1}{2} - k_0/k_j)^2}.$$

Some numerical values for the process described are of interest. Let $k_0 = 50$ vehicles/mile, $k_j = 200$ vehicles/mile, $u_f = 50$ miles/hr, $T = 0.1$ hr, (6 min).

The location of the shockwave when $t = 0$ is

$$x = - \frac{k_0}{k_j} u_f T = -1.25 \text{ miles,}$$

and $200 \times 1.25 = 250$ cars would be stopped. The starting wave would overtake the stopping wave at time

$$t_0 = \frac{T}{(k_j/k_0 - 1)} = \frac{0.1}{3} = 0.033 \text{ hr.}$$

The location would be at

$$x = -u_f t_0 = -50 \times 0.033 = -1.66 \text{ miles.}$$

The initial flow rate would be

$$q_0 = u_f \left(k_0 - \frac{k_0^2}{k_j} \right) = 50 \left(50 - \frac{2500}{200} \right) = 1875 \text{ vehicles/hr,}$$

and the waves carrying this flow would travel at a celerity of

$$c_0 = u_f \left(1 - \frac{2k_0}{k_j} \right) = 50 \left(1 - \frac{100}{200} \right) = +25 \text{ miles/hr.}$$

The equation of motion of the shockwave after the starting wave overtakes it is given by

$$x = 100(\tfrac{1}{4} - \tfrac{1}{2})t + 100(0.1 \times \tfrac{1}{4} \times \tfrac{3}{4}t)^{1/2},$$

$$x = -25t + 13.7t^{1/2}.$$

The speed of the shockwave is

$$c = \frac{dx}{dt} = -25 + 6.85t^{-1/2}.$$

Initially it is traveling upstream at the free speed $u_f = -50$, but it slows down and reverses, reaching zero when

$$t^{1/2} = \frac{6.85}{25} = 0.274,$$

$$t = 0.075 \text{ hr (4.5 min)}.$$

The shockwave will return to the point of its origin when $x = 0$ and

$$25t^{1/2} = 13.7,$$

$$t = \left(\frac{13.7}{25}\right)^2 = 0.302 \text{ hr (18 min)}.$$

Thus where the initial concentration is not too high, 50 vehicles/mile, a 6-min interruption will be cleared up within 18 min after the traffic starts up again, that is, three times the duration of the original interruption. If, however, the initial concentration has been nearly $k_j/2$, say, 90 vehicles/mile, the time to return to the initial conditions would be 100 times the duration of the original interruption.

Bottlenecks

Hydrodynamic theory is very useful in the understanding of traffic behavior at fixed and moving bottlenecks. Fixed bottlenecks were investigated by Lighthill and Whitham by considering a typical problem in which the capacity of a road is less in one section than it is in the sections upstream and downstream from it.

Figures 19 and 20 illustrate the phenomenon. Figure 19 gives the flow-concentration curve for the main road upstream and downstream from the bottleneck section and another flow-concentration curve for the bottleneck section itself. Points A and C are on the main-road curve and point B is on the bottleneck. As traffic increases up to point A on the first curve it reaches the maximum flow which can pass through the bottleneck. The wave carrying this flow cannot pass through the bottleneck and becomes stationary within it. Subsequent waves are reflected back upstream.

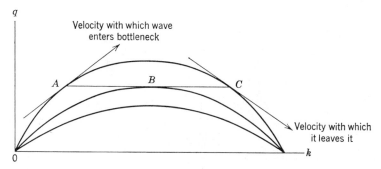

FIGURE 19. Illustration of the reflection of a wave from a bottleneck.

The effect on the vehicles in the traffic stream is that they approach the bottleneck initially at a speed given by the slope of the radius vector 0–A, but upon entering the bottleneck section they slow to a speed given by the slope of 0–B. When the waves are reflected upstream they travel upstream at a speed given by the tangent at point C. Vehicles encountering these reflected waves are forced to slow down to a crawl speed represented by the slope of the radius vector 0–C. The process is illustrated in Figure 20.

This phenomenon was investigated experimentally by Edie and Baverez.[48] Using seven pressure-sensitive tapes installed across a tunnel traffic lane in the bottleneck area (the beginning of the upgrade), data were obtained sufficient to construct the trajectories of vehicles at a time when the bottleneck became overloaded and to construct contours of speed values as shown in Figure 21.

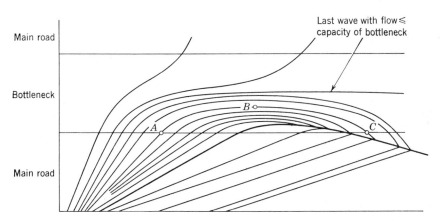

FIGURE 20. Formation of a shockwave in front of a traffic surge as it enters a bottleneck of inadequate capacity.

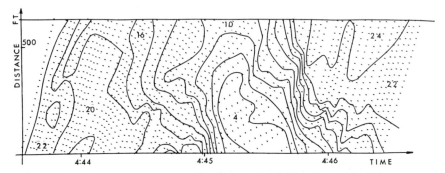

FIGURE 21. Trajectories and speed contours at a traffic bottleneck.

Using the trajectories as data, the mean values of the continuum variables q and k at numerous sampled points, located in the space and time area shown, were measured using the measurement method suggested by the principal author.[49] Averages were made in a rectangle having a length of 264 ft of space and a width of 18 sec of time. This amount of space and time is large enough to smooth out most random fluctuations and yet small enough to give a detailed picture of the dynamic changes taking place.

It will be observed in Figure 21 that the initial steady-state condition of 22–24 miles/hr rapidly deteriorates into unstable states of ever slower speeds. The slow speed condition propagates backward in space, passing out of the range of the observation at $4:45\frac{1}{2}$ p.m., following which the traffic stream returns to another steady state at the initial speed of about 22–24 miles/hr. All of this occurs within a period or about 3 min. It is understandable that a macroscopic theory, even though valid for dealing with many vehicles and long time periods, might not be applicable for describing such short-term transient phenomena.

What has taken place in this observation becomes fairly clear from a study of the flow and concentration contour maps in Figures 22 and 23, respectively. Referring to Figure 22, one observes a relatively high flow rate into this section at 4:44 p.m. and shortly thereafter, in the range of 1400–1800 vehicles/hour. These flow rates are unable to penetrate this section, and they are turned back to form peak contour regions of higher than usual flow rates. This initial state is followed by a minute or so of declining flow rates, the contour lines of which are first vertical but later tend to close on themselves downstream; upstream they tend to remain open-ended and to propagate, forming contour valleys of low flow. The lowest flow level contour shown encloses vehicles flowing at less than 500 vehicles/hr and less than 5 miles/hr.

With reference to Figure 23, one observes that changes in concentration occurred in a considerably different contour pattern from changes in flow. As would be expected from the continuity equation, $\partial q/\partial x + \partial k/\partial t = 0$, when

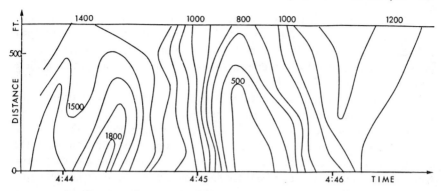

FIGURE 22. Flow contours at a traffic bottleneck, vehicles/hr.

the flow gradient in space is zero, the concentration gradient in time is zero; and the contours are therefore orthogonal. In this example, a critical condition arises when high concentrations build up to about 100 vehicles/mile and begin to propagate backwards. It is interesting that the contour lines of both q and k in Figures 22 and 23, respectively, approach slopes approximately the same as the vehicle trajectories at the beginning and end of the observation. In this steady state, changes in flow and concentration propagate at about the stream speed, which is a usual finding for relatively constant-speed, steady-state conditions. This result suggests that for this scale of measurement, small changes in flow may not propagate at a speed equal to the slope of the tangent to a steady-state q–k curve as suggested by the hydrodynamic wave theories of traffic flow. Instead, they are carried along at about stream speed or only slightly less than stream speed right up to saturation flows, at which level they suddenly reverse direction.

Leutzbach[50] also experimented with the usefulness of macroscopic theory with regard to bottlenecks. He observed that when at a given point x_0 the

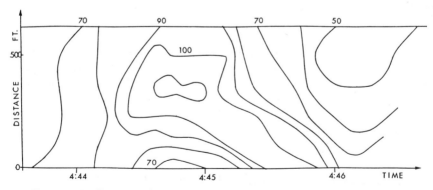

FIGURE 23. Concentration contours at a traffic bottleneck, vehicles/mile.

variation of concentration with time is known, and when the relationship between flow q and concentration k is known also from observations of the q–k curve, the theory describes the curve along which concentration is constant. Where such lines intersect, shockwaves occur. The speeds of these shockwaves can also be evaluated.

When the x–t plane has been filled with lines of equal concentration, the concentration can be read for any other point x_i, provided hydrodynamic theory is valid and the roadway is homogeneous with respect to the q–k curve. To test this, Leutzbach compared the observed concentration with the concentration predicted from the slope of the tangent to the q–k curve and found discrepancies. To describe these discrepancies numerically he compared the area under the k versus t curves for the observed k, measured as A, with the area under the theoretical curves of k versus t, measured as B, and the area between the two curves, measured as C. He found

$$a_1 = \frac{A - B}{A} = 12.7\% \text{ at the entering point,}$$

$$a_4 = \frac{A - B}{A} = 21.0\%, \text{ at the exiting point,}$$

$$b_1 = \frac{C}{A} = 19.1\% \text{ at the entering point,}$$

$$b_4 = \frac{C}{A} = 20.8\% \text{ at the exiting point,}$$

for points plotted as sliding 1-min values taken every 5 sec.

When data were combined into sliding 5-min values, he found that

$$a_1 = 9.8\%$$

and

$$a_4 = 10.8\%.$$

These results suggest no definite answer about how far fluid flow theory may be used in predicting the concentration at one point from knowledge of its concentration at another point and an empirically derived q–k curve assumed to apply at all points on the roadway. Thus, while hydrodynamic theory is extremely useful in describing qualitative changes in the traffic stream, its quantitative predictions have fairly limited accuracy. Nevertheless, it is useful to look at developments of the theory which seek to yield quantitative results from empirical data.

One useful example is given by that of a moving bottleneck created on a single-lane roadway by a slow-moving truck.

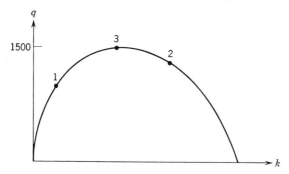

FIGURE 24. Traffic states (1) before a slow truck enters (2) behind the truck, and (3) at point of truck turn off on q–k curve.

Figures 24 and 25 illustrate the effects which may be predicted from hydrodynamic traffic theory about a slow-moving truck. Figure 24 shows the location of three traffic states involved in the problem. In state 1, there is a fluid traffic lane carrying a flow of $q_1 = 1000$ vehicles/hr at a concentration of $k_1 = 20$ vehicles/mile and a speed of $u_1 = 50$ miles/hr. A construction truck enters this roadway and travels at 12 miles/hr for 2 miles. Traffic piles up behind this truck to a concentration of $k_2 = 100$ vehicles/mile and while traveling at the speed of the truck, generates a flow of $12k_2 = 1200$ vehicles/hr. The end of the queue behind the truck to any fixed observer along the roadway brings a change in state from 2 to 1. This is a shockwave which has a celerity of

$$c_{21} = \frac{\Delta q}{\Delta k} = \frac{1200 - 1000}{100 - 20} = +2.5 \text{ miles/hr.}$$

This truck would take $\frac{1}{6}$th hour to travel the 2 miles. When he turns off, the wave 21 has traveled $2.5/6 = 0.4$ miles, and the queue is $2 - 0.4 = 1.6$ miles long. The queue contains $1.6 \times 100 = 160$ vehicles trapped behind the truck. At the point of the exit, the traffic speeds up to the optimum speed of point 3, $u_3 = 30$ miles/hr. The concentration drops to 50 vehicles/hr, and the flow increases to 1500 vehicles/mile. The acceleration wave which carries a change of state from state 2 to state 3 has a celerity of

$$c_{23} = \frac{1200 - 1500}{100 - 50} = -6 \text{ miles/hr.}$$

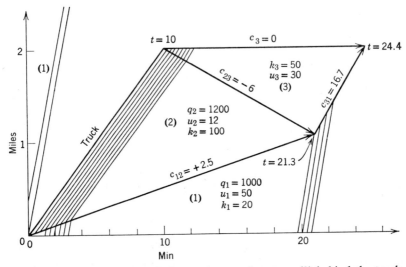

FIGURE 25. Traffic states (1) before a slow truck enters, (2) behind the truck, and (3) at point truck turns off on x–t diagram.

It reaches the end of the queue in an additional time of

$$t = \frac{1.6 \times 60}{6 + 2.5} = 11.3 \text{ min}$$

and at a location

$$x = 0.4 + \frac{2.5 \times 11.3}{60} = 0.88 \text{ miles.}$$

There is now a state 3 downstream and state 1 upstream from this point. This change of state propagates at a celerity of

$$c_{31} = \frac{1500 - 1000}{50 - 20} = +16.7 \text{ miles/hr.}$$

This wave requires and additional time of $1.12 \times 60/16.7 = 4.0$ min. The series of events and the times of their occurrences are the following:

Event	Time
Truck enters.	0
Truck leaves.	10 min
Queue is dissipated.	21.3 min
Traffic is normal at exit point.	25.3 min

In this solution, it should be noted there is an implicit assumption of starting shockwaves as well as slowing shockwaves, an assumption which differs from the Lighthill–Whitham theory and from Richards' theory.

The existence of acceleration shockwaves was suggested by Newell[51] as a means of explaining instabilities in dense traffic of the kind occurring in long vehicular tunnels, where the flow through the bottleneck section may be quite steady, yet the flow upstream from the bottleneck may be characterized as "stop-and-go" driving. In such tunnels one may observe a disturbance originate near the bottleneck location every 4–6 min, and as the disturbance propagates from one car to the next, it becomes amplified until some car comes to a complete stop. When an acceleration shock reaches the car it will start up again, possibly being forced to stop a second time by the next disturbance. Newell explains the cause of such acceleration shockwaves as lying in the existence of two different q–k curves, or two different speed-spacing curves, one followed when a car is accelerating, the other when decelerating. When doing neither, cars tend to drift between the two curves.

Instability arises in Newell's theory because waves for acceleration and deceleration do not necessarily propagate at the same speed. In a stable continuum theory, fluctuations are always dissipated. If a car accelerates a little and then returns to its original speed, it sends out two groups of waves. The acceleration waves expand, and the deceleration waves form a shock (see Richards' theory, Figures 17 and 18); but the acceleration waves have a range of speeds that includes the shock speed. This means that the shock always overtakes at least part of the expanding wave, and they proceed to annihilate each other. In Newell's theory the two types of waves do not necessarily travel at the same speed. If the wave first created travels faster than the opposite-type wave that follows it, the latter will never catch up, and the disturbance will become permanently embedded in an expanding region between the two waves.

There is considerable experimental evidence to support Newell's theory of asymmetry between acceleration and deceleration. Forbes et al.[52] reported slower response of drivers to acceleration than to deceleration. Herman and Potts[53] reported measurements that gave slightly different values for acceleration and deceleration. Daou[16] found significantly larger spacing between cars accelerating at a given speed than between cars decelerating. In addition, there have been numerous sets of trajectories of cars taken from aerial photographs that show the existence of accelerating shock waves.

Mathematical Treatment—Quantitative Models

Thus far, treatment of continuum theory has been limited to largely qualitative aspects or graphical solutions. We now consider various quantitative models in mathematical form.

Before the development of hydrodynamic theory of traffic flow, some specific quantitative models had been suggested by traffic engineers and researchers for the relationships between flow and concentration or speed and concentration. These were generally linear models in which, for example, the speed of the traffic stream was assumed to be a linear function of the concentration:

$$u = u_f\left(1 - \frac{k}{k_j}\right),$$

which leads to a parabola for the q–k curve:

$$q = u_f\left(k - \frac{k^2}{k_j}\right).$$

Another model assumed that the speed was a linear function of the flow, that is,

$$u = u_f\left(1 - \frac{q}{2q_m}\right),$$

where $u \geq u_f/2$.

A third model is based on the Uniform Vehicle Code of the state of California, which states that a car should keep a spacing of a car length behind the car ahead for each 10 miles/hr of speed; that is,

$$s = \frac{u}{a} + s_j,$$

where s = mean spacing between vehicles, rear to rear, s_j = jam spacing, and a = a constant = $10/L$, where L is car length. The following relationships then obtain:

$$u = a(s - s_j) = a\left(\frac{1}{k} - \frac{1}{k_j}\right),$$

$$q = a\left(1 - \frac{k}{k_j}\right),$$

$$a = q_m,$$

where q_m is the maximum flow which is approached as $u \to \infty$.

The first model for the q–k curve derived entirely from hydrodynamic theory was that proposed by Greenberg,[54] who assumed the equation of motion of a one-dimensional fluid and the continuity equation. The assumed equation of motion is

$$\frac{du}{dt} = \frac{-a^2}{k}\frac{\partial k}{\partial x}, \tag{24}$$

which states that the acceleration of the traffic stream, du/dt, is inversely proportional to the concentration k and directly proportional to the concentration gradient $\partial k/\partial x$, with a constant of proportionality of $-a^2$. If the speed is given as a function of location and time, that is, $u = u(x, t)$, the equation of motion becomes

$$\frac{\partial u}{\partial t} + u\frac{\partial u}{\partial x} + (a^2/k)\frac{\partial k}{\partial x} = 0.$$

If $q = uk$,

$$\frac{\partial q}{\partial x} = u\frac{\partial k}{\partial x} + k\frac{\partial u}{\partial x},$$

and substituting for $\partial q/\partial x$ in the continuity equation, it becomes

$$\frac{\partial k}{\partial t} + u\frac{\partial k}{\partial x} + k\frac{\partial u}{\partial x} = 0. \tag{25}$$

If it is now assumed that the speed of the traffic stream at a point is only a function of concentration, that is, $u = u(k)$, we can find

$$\frac{\partial u}{\partial x} = \frac{du}{dk}\frac{\partial k}{\partial x}, \qquad \frac{\partial u}{\partial t} = \frac{du}{dk}\frac{\partial k}{\partial t}.$$

Substituting these into equations and solving for du/dk, we have

$$\frac{du}{dk} = -\frac{a}{k}.$$

Integrating, and using initial conditions $u = 0$ when $k = k_j$ yields

$$u = a \ln\left(\frac{k_j}{k}\right)$$

$$q = ku = ak \ln\left(\frac{k_j}{k}\right).$$

Differentiating and setting dq/dk to zero gives $k_j/k = e$ as the concentration at maximum flow, and

$$a = u_m,$$

$$q = u_m k \ln\left(\frac{k_j}{k}\right). \tag{26}$$

Equation 26 gives a good fit to experimental data taken in the Lincoln Tunnel in New York, as shown in Figure 26.

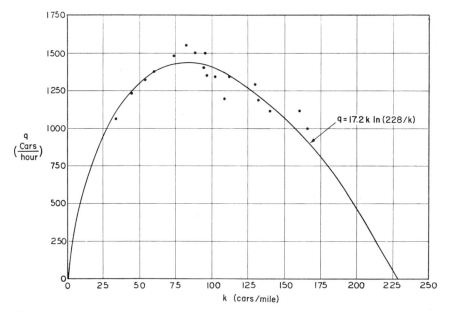

FIGURE 26. Fit of Greenberg's model to experimental data from the Lincoln Tunnel.

This quantitative model for the q–k curve is equivalent to a model in which the speed of propagation of disturbance in the traffic stream relative to the stream itself c_r is a constant:

$$c_r = a = u_m.$$

Physically, this may be interpreted as saying that a signal propagates from one vehicle to the next at a constant speed, and the response time is proportional to the spacing between them.

The speed of a disturbance relative to the roadway, the wave celerity, is

$$c = u - u_m,$$

and the wave stands still when the speed is that for maximum flow.

There are a number of such models which may be derived directly from different assumptions about the behavior of c_r, namely, that it is proportional to spacing, concentration, or flow. The results of making such assumptions are tabulated in Table 5. It will be noted that two expressions are given for the acceleration of the stream, du/dt. One is in the form of an equation of motion of a fluid as was used by Greenberg. The other results in differentiating the steady-state expression for speed, u, with respect to time. These relate the acceleration to the relative speed of vehicles in the stream relative to two

TABLE 5

RELATIONSHIPS OF VARIOUS CONTINUUM MODELS[a]

Variable	California	Greenshields–Richards	Greenberg–Herman	Underwood–Edie
$q =$	$q_m\left(1 - \dfrac{k}{k_j}\right)$	$u_f\left(k - \dfrac{k^2}{k_j}\right)$	$u_m k \ln\left(\dfrac{k_j}{k}\right)$	$k_m u \ln\left(\dfrac{u_f}{u}\right)$
$u =$	$q_m\left(\dfrac{1}{k} - \dfrac{1}{k_j}\right)$	$u_f\left(1 - \dfrac{k}{k_j}\right)$	$u_m \ln\left(\dfrac{k_j}{k}\right)$	$u_f\, e^{-k/k_m}$
$c =$	$\dfrac{-q_m}{k_j}$	$u_f\left(1 - \dfrac{2k}{k_j}\right)$	$u - u_m$	$u\left(1 - \dfrac{k}{k_m}\right)$
$c_r =$	$\dfrac{q_m}{k} = as$	$\dfrac{u_f k}{k_j} = ak$	$u_m = a$	$\dfrac{q}{k_m} = aq$
$\dfrac{du}{dt} =$	$q_m \dfrac{ds}{dt}$	$\left(\dfrac{u_f s_j}{s^2}\right)\dfrac{ds}{dt}$	$\left(\dfrac{u_m}{s}\right)\dfrac{ds}{dt}$	$\left(\dfrac{s_m u}{s^2}\right)\dfrac{ds}{dt}$
$\dfrac{du}{dt} =$	$\left(\dfrac{-a^2}{k^3}\right)\dfrac{\partial k}{\partial x}$	$(-a^2 k)\dfrac{\partial k}{\partial x}$	$\left(\dfrac{-a^2}{k}\right)\dfrac{\partial k}{\partial x}$	$(-a^2 uq)\dfrac{\partial k}{\partial x}$
where $a =$	q_m	$\dfrac{u_f}{k_j}$	u_m	s_m

[a] q = flow (vehicles/hr); k = concentration (vehicles/mile); k_j = jam concentration (vehicles/mile); k_m = optimum concentration (vehicles/mile); u = traffic speed (miles/hr); u_f = free traffic speed (miles/hr); u_m = optimum traffic speed (miles/hr); s = spacing (miles); s_m = optimum spacing (miles); du/dt = acceleration (miles/hr²); ds/dt = relative speed (miles/hr); $\partial k/\partial x$ = concentration gradient.

adjacent vehicles. In Section V, these expressions will be shown to relate directly to the microscopic theory of car following.

Some Further Mathematical Considerations

If a traffic stream can be treated mathematically as a compressible fluid, it should be feasible to start with some initial conditions and functional relationships and predict subsequent values of q, k, and u in space and time. The papers of Lighthill and Whitham and that of Richards, have shown how to do this largely by following the progress of disturbances largely graphically. While these methods are perhaps the easiest and most suitable for practical situations, algebraic solutions are also of interest.

Richards showed that the solution to the differential equations of motion and continuity were of the form

$$k = f(x - ct),$$

where f is some arbitrary function of the argument $(x - ct)$ and c is the speed of propagation of a small disturbance. It can be shown that the concentration remains constant along the characteristic $c = dx/dt$, and when the flow is a function of k only, the flow is also constant along $c = dx/dt$. If

$$q = f(k),$$

$$\frac{\partial q}{\partial x} = \left(\frac{\partial k}{\partial x}\right)\frac{dq}{dk},$$

and as previously shown $dq/dk = c = dx/dt$. Substituting this in the continuity equation yields

$$\left(\frac{\partial k}{\partial t}\right) dt + \left(\frac{\partial k}{\partial x}\right) dx = 0,$$

and if k is a function of x and t,

$$k = k(x, t),$$

$$dk = \left(\frac{\partial k}{\partial x}\right) dx + \left(\frac{\partial k}{\partial t}\right) dt,$$

and therefore

$$dk = 0 \text{ along } c = \frac{dx}{dt}.$$

In order to find $k = k(x, t)$ we look for a suitable arbitrary function f. As an example, a simple linear equation of state will be assumed:

$$q = q_m\left(1 - \frac{k}{k_j}\right),$$

where

$$\frac{dx}{dt} = \frac{dq}{dk} = c = -\frac{q_m}{k_j},$$

$$k = f\left[x + \left(\frac{q_m}{k_j}\right)t\right].$$

Now if the initial conditions are

$$k(x, 0) = k_0(1 + ax),$$

where the concentration increases linearly with x from a value of k_0 at the origin at a rate a per mile. A suitable arbitrary function would then be

$$k(x, t) = k_0 + ak_0\left[x + \left(\frac{q_m}{k_j}\right)t\right],$$

$$q(x, t) = \frac{q_m}{k_j}\left[k_j - k_0 - ak_0\left(x + \frac{q_m}{k_j}t\right)\right].$$

These are solutions to the continuity equation since

$$\frac{\partial q}{\partial x} = -\left(\frac{q_m}{k_j}\right)ak_0,$$

$$\frac{\partial k}{\partial t} = +\left(\frac{q_m}{k_j}\right)ak_0.$$

If the initial conditions are

$$k(x, 0) = k_0[1 - \exp(-bx)],$$

a suitable solution would be

$$k(x, t) = k_0 - k_0 \exp\left(-bx - \frac{bq_m t}{k_j}\right).$$

Where the normal equation of continuity does not hold, one can also derive explicit solutions for $f(x - ct)$. We will consider a roadway on which there are entering and exit ramps such that

$$\frac{\partial k}{\partial t} + \frac{\partial(ku)}{\partial x} = D(x, t),$$

where the function $D(x, t)$ describes the influx at point x at time t. We assume that traffic increases exponentially along the roadway. Thus

$$\frac{\partial k}{\partial t} - \left(\frac{q_m}{k_j}\right)\frac{\partial k}{\partial t} = A \exp(-bx).$$

We let $k = f(x, t) + g(x)$, and can by standard methods solve for

$$k(x, t) = c_1 \exp\left(-bx - \frac{bq_m t}{k_j}\right) + \left(\frac{k_j}{q_m}\right)Ab \exp(-bx) + c_2,$$

where c_1 and c_2 are constants of integration, which can be evaluated from initial conditions. If we assume the initial speed distribution is linear with x, namely,

$$u(x, 0) = u_0 + rx,$$

the constants are

$$c_1 = -\left(\frac{k_j}{q_m}\right)\left(\frac{A}{b}\right),$$

$$c_2 = \frac{q_m}{(u_0 + rx + q_m/k_j)}.$$

It can be seen that even with simple mathematical assumptions, an explicit analytical solution of traffic flow problems becomes rather complicated. It is therefore necessary to resort to computer solutions in any but the simplest situations. Even higher-order complications are introduced when the flow at a point is not a function of just the concentration, but is also a function of the concentration gradient and time rate of change; that is,

$$q = q(k, k_x, k_t).$$

When

$$k_x = \frac{\partial k}{\partial x}, \qquad k_t = \frac{\partial k}{\partial t}.$$

The continuity equation then becomes

$$k_t = ck_x + pk_{xt} + nk_{xx} = 0,$$

where

$$k_{xt} = \frac{\partial^2 k}{\partial x\,\partial t}, \qquad k_{xx} = \frac{\partial^2 k}{\partial x^2},$$

$$n = \frac{\partial q}{\partial k_x}, \qquad p = \frac{\partial q}{\partial k_t}.$$

Experiments and Applications of Macroscopic Theory

Many experiments have been made to test and apply the concept of a hydrodynamic analogy for vehicular traffic flow. Some have been mentioned already. A few more will be discussed, especially the pioneering experiments made in tunnels operated by the Port of New York Authority.

Edie and Foote[55-58] have made many determinations of the flow-concentration relationships and wave propagations in the tunnels. The initial experiments were for the purpose of determining a capacity profile of the tunnel to identify bottleneck locations. In addition to making mass-flow observations at numerous locations to develop flow-concentration relationships, numerous controlled experiments were run with platoons traveling near the optimal speeds, 20–25 miles/hr.

Figures 26 and 27 show a typical curve fitted to Greenberg's model over the entire range of concentration and the same data fit to two models, the

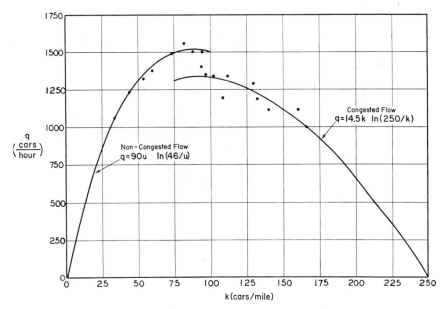

FIGURE 27. Fit of two models (Greenberg) and (Edie–Underwood) to Lincoln Tunnel data.

Edie–Underwood model for noncongested traffic, less than 90 vehicles/mile of concentration, and the Greenberg model for congested traffic at greater concentrations. The two-model fit suggests that the curve has a discontinuity around a $k = 90$ vehicles/mile. The one-model fit gives a curve of $q = 17.2\,k \ln(228/k)$ and the two-model fit gives curves of

$$q = 90u \ln\left(\frac{46}{u}\right)$$

and

$$q = 14.5 \ln\left(\frac{250}{k}\right).$$

The parameters of the first are $k_m = 90$ vehicles/mile, $u_f = 46$ miles/hr and those of the second are $u_m = 14.5$ miles/hr and $k_j = 250$ vehicles/mile.

Bottlenecks were also located by means of mass-flow, time-sequence observations which revealed the origins and the celerities of shockwaves, which were found to originate at the bottleneck locations found by the flow-concentration experiments. Figure 28 gives the results of one of the wave experiments. Observations were made at seven locations, A through G, from the tunnel entrance to the exit. When one studies the time-sequence curves

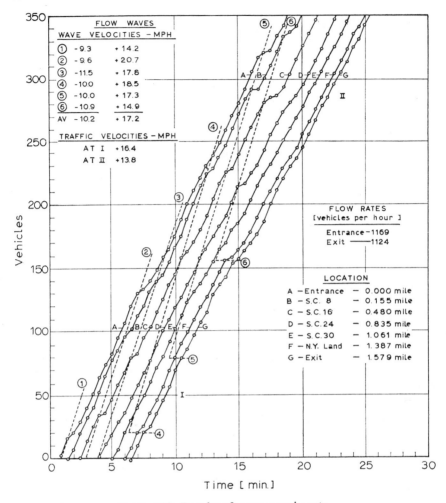

FIGURE 28. Results of wave experiments.

for several observers, the propagation of shockwaves becomes evident. In the figure they are shown by dashed lines. The space and time origins of these waves are apparent, and the velocities can readily be measured at the values shown for the six shockwaves observed. For comparison with these values, which average -10.2 miles/hr, kinematic wave theory would estimate the backward waves at about 10 miles/hr for the conditions

$$c = \frac{\Delta q}{\Delta k} = \frac{1400 - 0}{70 - 210} = -10 \text{ miles/hr}.$$

Also one could estimate the forward wave speeds using a Δq of 1200 vehicles/hr and a Δk of 70 vehicles/mile to be

$$c = \frac{\Delta q}{\Delta k} = \frac{1200}{70} = +17 \text{ miles/hr}.$$

It will further be noted from the figure that waves 4–6 appear to originate between observers E and F. This is the region in the tunnel at the beginning of the upgrade.

In the platoon experiments, groups of 11 vehicles were sent through the tunnel near optimal speeds, and the time intervals between the first and last vehicles were measured at many points in the tunnel. The capacity profiles obtained from the platoon experiments and the other (mass-flow) experiments are shown in Figure 29. The results show a difference of 300 vehicles/hr in the minimum average flow rate as compared to continuous flow. Such a difference leads to the belief that the interaction between platoons of vehicles causes the loss of flow observed when the traffic state changes from non-congested behavior to congested behavior. Greater flows might be achieved by breaking these interactions.

This hypothesis was confirmed by what were termed *gap experiments*, during which gaps were introduced into the traffic stream at the entrance to

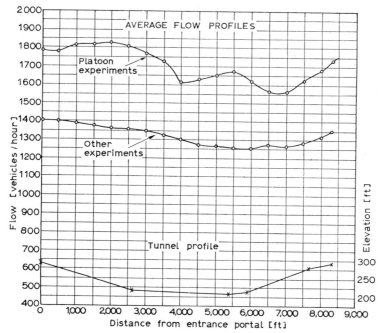

FIGURE 29. Capacity profiles.

the tunnel when as many as 44 vehicles had entered one lane within 2 min. The gap experiments yielded an average increase of flow of 70 vehicles/hr. These results were reported by Greenberg and Daou.[59] Figure 30, a graph of the time-sequence observations by four observers during gap experiments, illustrates a particular effective use of "gapping" during a 20-min period. The rate of flow was increased from 1200 to 1360 vehicles/hr. Gaps of 45 sec (SG) were introduced at the entrance on the appearance of slow-down shockwaves such as 1 and 2. Small gaps (G) were introduced at 2-min intervals whenever 44 vehicles had entered. Experience with this control system showed, however, that gaps based on a fixed number of entering vehicles in 2 min did not always work satisfactorily. Some of the gaps were not long enough, and sometimes they were too long or unnecessary.

This then led to a feedback system of control of the traffic through the Lincoln Tunnel in New York City, using a computer. This system, in its most advanced form, is described elsewhere under traffic control.

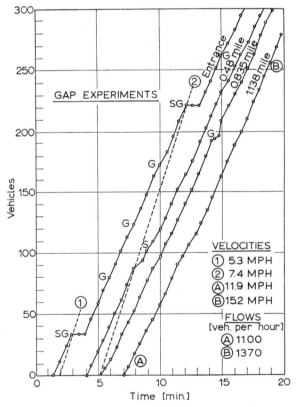

FIGURE 30. Results of gap experiments.

Some interesting experiments on the speed-flow relationship were conducted at the Road Research Laboratory in England in 1962 and 1963 and reported by Wardrop.[60] These were run on four circular and one straight track. In most of the experiments on the circles, a fixed number of vehicles were assembled on the central area of the track in a stationary queue, and the queue was then released. In some tests the vehicles were formed into a single platoon. The leading driver was instructed to travel at a fixed speed and all the other drivers were instructed to follow the vehicle ahead as closely as was comfortable.

Principal results are given in Figure 31 in the form of speed-concentration curves. All of the curves except that for the 56-ft radius are similar and converge into a single curve at high concentrations, above 80 vehicles/mile. The free speeds (zero concentration) for the circular roads corresponded to a centrifugal force of about $\frac{1}{3}$ g.

Franklin[61] made some further analyses of the data with some additional interesting results. He found at least two distinct regimes of flow; for example, on the 106-ft-radius track there was one relationship for concentrations below 100 vehicles/mile and another for concentrations greater than this.

The reason for this is a simple but important one. If the track is congested

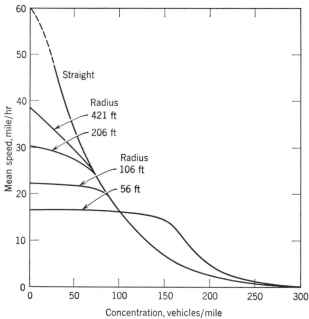

FIGURE 31. Speed–concentration curves from test tracks.

and one or two vehicles are removed, the remaining vehicles are able to increase their speeds because they have, on the average, more space ahead of them. However, if this is continued, the speed of the vehicles on the track will eventually rise to a value beyond which the drivers are not willing to go because of centrifugal force or safety considerations. Further reduction in the number of vehicles will not cause an increase in speed; instead, gaps form and the vehicles break up into platoons. Since there is no increase in speed by further decreases in concentration, the flow falls off linearly with reduction in the concentration.

The flow-concentration curves found by Franklin supported a model in which the flow-concentration curve was triangular. The flow increased linearly with concentration up to a critical value at maximum flow and then decreased linearly with concentration at higher and higher values, reaching zero at the jam concentration.

Forbes and Simpson[62] made aerial photos of peak-hour traffic on the John Lodge Expressway in Detroit. Using space–time diagrams to show the results (namely, trajectories of all the vehicles observed), they were able to follow the progress of deceleration waves and to measure spacings and time headways before and after the waves. The results reflect the interaction between driver response time, speed, and headways. Table 6 below gives some of these for waves producing stoppages.

The results suggest that where average propagation response time is used as a driver sensitivity coefficient, it will not remain constant in all traffic situations. Deceleration wave response time approaches true driver sensitivity when time headways are very short. With longer headways, response includes an additional time lag. Longer time headways occur under conditions of driver uncertainty and anticipation of congestion ahead. Also, there is a lengthening of headways after decelerations and a slow return to a prior shorter headway.

TABLE 6

TRAFFIC AND WAVE SPEEDS AND RESPONSE TIMES

Traffic Speeds, ft/sec			Wave Speeds, ft/sec		Response Times, sec	
Before	During	After	Dec.	Acc.	Dec.	Acc.
43.0	2.7	30.2	11.9	2.5	1.96	2.00
41.1	1.0	43.0	15.0	12.5	1.88	2.40
40.0	2.3	39.9	17.5	16.9	1.42	1.52
35.1	4.5	37.8	12.5	16.9	2.29	2.13
55.0	1.6	45.5	20.0	17.5	1.26	1.19

Foster[63] measured the dissipation of a traffic shock front when a platoon of traffic is released on the green signal at a pedestrian-controlled traffic light and used the results to study the applicability of kinematic wave theory to single-lane traffic flow. Essentially the method used allowed plotting the trajectories on a space–time diagram of vehicles stopped by the signal as they started up.

A principal result of the experiments was to show that starting waves drawn through points on the vehicle trajectories with the same speeds were approximated by straight lines, as would be the case when flow at a point is a function of concentration only. Spacings and headways could be readily determined by measurements made along these characteristic lines, including their distributions.

Another principal result was to show that the traffic states of vehicles while accelerating could be fit with models similar to those that fit steady-state traffic conditions. Two of the models that fit with correlation coefficients close to 1.0 were the Greenshields model:

$$u = 29.7\left(1 - \frac{k}{181}\right),$$

where

$$u_f = 29.7 \text{ miles/hr} \quad \text{and} \quad k_j = 181 \text{ vehicles/mile}$$

and the Greenberg model:

$$u = 19.9 \ln\left(\frac{194.5}{k}\right),$$

where

$$u_m = 19.9 \text{ miles/hr} \quad \text{and} \quad k_j = 194.5 \text{ vehicles/mile}.$$

An interesting application of macroscopic theory was made by Smeed[64] in showing that because of the flow-concentration relationship for roads, there were circumstances in which vehicles could reach their destinations sooner by starting later. Smeed showed this for travel to work in an urban environment. Here we will show it for a single roadway.

Since the speed of travel on a roadway usually decreases as the flow increases, there may be an advantage in extending the period over which vehicles are permitted to enter a roadway in order to increase their speed of travel. If the speed of travel is increased sufficiently to more than compensate for the delayed entry, all vehicles using the roadway may arrive earlier at the destination by starting later.

If n vehicles enter in an access time t and travel a distance l at a speed v, the last vehicle will arrive at

$$T_n = t + \frac{l}{v}.$$

The minimum value of T_n occurs when

$$\frac{dT_n}{dt} = 1 - \left(\frac{l}{v^2}\right)\frac{dv}{dt} = 0.$$

Smeed[64] has shown how to determine these minima for vehicles entering a central business district of a large city. It is interesting from the standpoint of traffic flow theory also to look at circumstances on a single roadway for which vehicles can arrive earlier at their destination by starting later.

The simplest relationship, and a reasonably realistic one, to assume between v and the flow rate q is linear where

$$v = v_f\left(1 - \frac{q}{2q_m}\right),$$

and q_m is the maximum flow which is assumed to occur at one-half the free speed v_f. Since

$$q = \frac{n}{t},$$

$$\frac{dv}{dt} = \frac{v_f n}{2q_m t_n^2}.$$

Therefore, at minimum T_n,

$$\frac{v_f n l}{2q_m t_n^2} = v_f^2\left(1 - \frac{n}{2q_m t_n}\right)^2$$

and solving

$$t_n = \frac{n}{2q_m} + \left(\frac{nl}{2q_m v_f}\right)^{1/2},$$

for

$$l \geq \frac{n v_f}{2q_m}$$

give the entry time of the nth vehicle, which minimizes his overall time T_n. For the rth vehicle, $1 \leq r \leq n$, the minimum overall time is

$$T_r = \left(\frac{r}{n}\right)t_n' + \frac{l}{v},$$

and the average for all vehicles is

$$T = \frac{t_n'}{2} + \frac{l}{v},$$

and where t'_n is still the entry time of the nth vehicle, the optimum for all n vehicles occurs when

$$t'_n = \frac{n}{2q_m} + \left(\frac{nl}{q_m v_f}\right)^{1/2},$$

for

$$l \geq \frac{nv_f}{4q_m}.$$

Some numerical examples are of interest to show how the optimum flow rate q is related to the maximum flow rate for different road lengths.

Table 7 summarizes results obtained assuming the following values:

$$v_f = 50 \text{ miles/hr},$$

$$q_m = 1500 \text{ vehicles/hr},$$

$$n = 1500 \text{ vehicles}.$$

TABLE 7

OPTIMAL TIME PERIODS FOR ENTERING 1500 VEHICLES

Length of Roadway	Optimal Conditions for nth Vehicle			Optimal Conditions for Average Vehicle		
l, miles	t, hr	q, vehicles/hr	v, miles/hr	t, hr	q, vehicles/hr	v, miles/hr
12.5	1	1500	25	1	1500	25
25	1	1500	25	1.21	1260	29.4
50	1.21	1260	29.4	1.50	1000	33.3
100	1.50	1000	33.3	1.91	782	37.1

Table 8 shows the total of access time t plus travel time l/v for input to the roadway at capacity flow and input spread over the optimal access time. For capacity flow total time is T_m, for optimal conditions T. The saving in total time is $T_m - T = \Delta T$.

Summary of Macroscopic Theory

It is clear that macroscopic theory will always be useful in studying traffic behavior under conditions of heavy traffic. It has already provided a sound basis for understanding and predicting the behavior of groups of vehicles, especially in a qualitative way. Some of the quantitative models have also been found to hold up well in real traffic situations. A weakness of macroscopic theory is, of course, that it does not incorporate driver, vehicle, and

TABLE 8

TIME SAVINGS FOR DELAYED ENTRANCE

Length of Roadway,	For nth Vehicle Total Time & Saving, hr			For Average Vehicle Total Time & Saving, hr		
l, miles	T_m	T	ΔT	T_m	T	ΔT
25	1.5	1.5	0	1.50	1.46	0.04
50	3.0	2.9	0.1	2.50	2.25	0.25
100	5.0	4.5	0.5	4.50	3.65	0.85

roadway parameters in an explicit way. This is not true with microscopic theory for heavy traffic flow, which is covered in the last section of this chapter.

V. MICROSCOPIC MODELS

Microscopic models of traffic flow involve a description of the behavior of single vehicles. The most important models yet developed are known as car-following models. These provide the basic formulations for follow-the-leader traffic theory through which certain characteristics of a stream of interacting vehicles can be described by means of driver–vehicle parameters. Such models are more fundamental to traffic flow theory than hydrodynamic models, which simply look at a traffic stream as an analog of a continuous fluid and use fluid rather than driver–vehicle parameters.

Car-Following Models

Car-following models provide quantitative values over time of the acceleration of one vehicle following another when the leading vehicle is changing its speed over time. If this can be done with two vehicles, the behavior of a third vehicle following the second can be determined by the same model provided that there is no direct response of the third vehicle to the behavior of the first. However, this, too, can be taken into account if it exists. Under such conditions the dynamics of an entire single lane of traffic can be generated from the one model when the appropriate parameters of each driver and vehicle are known. We can predict the behavior of the second vehicle from the first, the third vehicle from the second, fourth from the third, and so on. However, in order to do this the vehicles must all be interacting, each one with the one or more ahead. The theory is rigorous only for heavy traffic, in which there is virtually no passing or lane-changing on multilane highways.

In single lanes, the theory seems applicable to moderate traffic levels in describing flow-concentration relationships from driver–vehicle parameters.

Car-following theory was first proposed by Reuschel[65] and Pipes[66] in the form of differential-difference equations. Their work has been greatly extended theoretically by Herman and others[67–75] who even more importantly devised experimental methods and equipment with which to test the theory and to measure the parameters. The description which follows is taken almost entirely from their published papers.

The car-following model of Herman and others is based on the concept that a driver accelerates, or decelerates, his vehicle in response to a stimulus or several stimuli he gets from his driving environment, which in the simplest form is given by his relationship or, rather, changes in his relationship to the vehicle ahead. The general form is

$$\text{response (now)} = \text{sensitivity} \times \text{stimulus (previously)}.$$

As indicated above, the response is made at some time lag after the stimulus occurs: This accounts for the reaction time on the part of the driver, required for him to perceive, interpret, evaluate, and act, and a mechanical lag in the vehicle before it can respond to the driver's commands.

The question arises as to which stimulus in the car-following situation is most important in causing a driver to accelerate or decelerate and in affecting the magnitude of their values. We might assume the stimulus to be the deviation from a desired spacing, the desired spacing being a function of the speed—the relative speed, absolute speed, and so on.

A car-following law (for example, one that states that the acceleration of a car at a delayed time is directly proportional to the relative speed of the car with respect to the one ahead) is a grossly simplified description of a very complicated response to many stimuli. The response–stimulus description would in practice be a highly complicated functional of the dynamical properties of the cars and of the psychological and physical characteristics of the drivers.

It is surprising that the car-following phenomenon can be so well approximated by a simple continuous differential-difference equation. The reason for this is, of course, that a driver, in seeking to avoid accidents and get to his destination quickly, cooperates with other drivers by using a simple average response to a limited number of stimuli.

Two of the most important traffic problems today are accidents and increased traffic flow in the face of an increasing concentration of traffic. Car-following theory sheds light on both problem areas. Rear-end and chain collisions can result from local and asymptotic instability in a line of moving cars in their response to a perturbation introduced by a lead car. Traffic flow-concentration relationships can be derived from car-following laws.

Before proceeding to the details of recent research on car-following theory, the importance of simple models that can be tested experimentally cannot be overemphasized. A simple model with its exact behavior as a guide can be used to make predictions that, when tested by experiments, will give new insights into traffic dynamics.

We begin with the simplest stimulus–response relationship, known as the linear car-following model. In this relationship it is assumed that acceleration (response) is directly proportional to the relative velocity (stimulus). Mathematically this is expressed as

$$\ddot{x}_{n+1}(t + T) = a[\dot{x}_n(t) - \dot{x}_{n+1}(t)], \qquad (27)$$

where n is the number of the lead car and $n + 1$ that of the follower, T is the time lag, and a is the coefficient of proportionality, in this case, a constant.

At instants when the lead car is going faster than the follower, the latter applies an accelerating force and vice versa. In the simplest case it is assumed that the sensitivities are identical for acceleration and deceleration. The stability of a line of traffic depends on whether a local fluctuation in speed is damped out or amplified as it propagates down the line of cars. Asymptotic stability conditions depend on equilibrium spacing and speed. If the spacing is small, a collision will occur a few cars back of the disturbing car.

Since this simplest law is linear, the stability question can be investigated in terms of the Fourier components of the driving function $\dot{x}_0(t)$, that is, the speed changes of the lead car. If we assume that the driving function is monochromatic with frequency, then

$$\dot{x}_0(t) = \exp(i\omega t).$$

An arbitrary driving function can be expressed as a linear combination of monochromatic components by the usual Fourier analysis. Then by substituting

$$\dot{x}_n(t) = f_n \exp(i\omega t)$$

into the equation, we obtain

$$f_n = \left[1 + \left(\frac{i}{a}\right)\right] \omega \exp(i\omega T)^{-n} f_0.$$

The amplitude factor decreases with a in

$$\frac{\omega}{a^2} > \frac{\sin \omega T}{2a}.$$

As $\omega \to 0$, the most critical condition arises and we find for asymptotic stability the sensitivity times the time lag must be less than $\frac{1}{2}$.

$$aT < \tfrac{1}{2}.$$

If only two cars are involved instead of a line of cars, one would be interested in whether the second car would amplify the signal to such an extent as to oscillate with increasing amplitude and eventually to cause a collision, or whether it would oscillate stably, oscillate with damping, or not oscillate at all.

For the spacing between two cars, these conditions arise for the following relationships of aT, which we shall identify as the dimensionless parameter C:

If $C > \pi/2$ (1.57), the spacing is oscillatory with increasing amplitude.
If $C = \pi/2$, oscillatory with constant amplitude.
If $1/e < C < \pi/2$, oscillatory with damped amplitude.
If $C < 1/e$, nonoscillatory and damped.

Figures 32–35 show results which may be derived from the car-following theory.

Figure 32 shows the effect on car 2 from a deceleration–acceleration maneuver by car 1, when car 2 is operating on the limit of damped response, where $C = aT = e^{-1} = 0.368$, $T = 1.5$ sec. The original acceleration and velocity pulses are smoothed out and damped by car 2, which returns to the original spacing of 70 ft about 12 sec after the maneuver began.

Figure 33 illustrates the effects on spacing of critical values of C above this level. At $C = 0.50$ and $C = 0.80$ there is oscillatory but damped response; at $C = \pi/2 = 1.57$ there is oscillatory response with constant amplitude; and at $C > \pi/2 = 1.60$ there is oscillatory response undamped and therefore unstable.

Figure 34 illustrates the behavior of a line of light cars at various values of C. When $C = e^{-1} = 0.368$, the response is stable and well damped; when $C = 0.5$, which is the limit of stability, the responses are very slightly damped; and when $C = 0.75$, responses are amplified and the condition is unstable.

Figure 35 illustrates the ultimate consequences of instability. Here the trajectories of nine cars are shown in a coordinate system moving at the initial speed of the cars. It can be observed that because of the unstable manner of driving, the disturbance begun by car 1 is amplified by each successive car, and the result is a collision between the cars 7 and 8. The values assumed are $C = 0.8$, $T = 2$ sec, and the initial spacing is 40 ft.

In order to obtain statistical estimates of certain functions and parameters it is necessary to design and conduct experimental studies of driver–car performance. Herman and co-workers were concerned with measuring the spacing between a lead car at location $x_l(t)$ and a follower car at $x_f(t)$ at time t, namely, $s(t) = x_l(t) - x_f(t)$; and the speeds, $u_l(t)$, $u_f(t)$, relative speed $u_l(t) - u_f(t)$, and the acceleration of the follower car, $a_f(t)$.

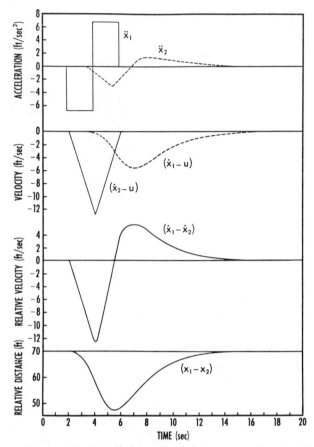

FIGURE 32. Detailed motion of two cars showing the effect of a fluctuation in the acceleration of the lead car. The second car follows the first with relative speed control and a time lag $T = 1.5$ sec and $C = aT = e^{-1} = 0.368$, the limiting value for local stability.

To measure the spacing and relative speed, a reel of piano wire and a power unit was designed and installed on the follower car. Several hundred feet of wire were wound on the reel, and the end of the wire was fastened to the rear bumper of the lead car. A constant wire tension was maintained by means of a slipping friction clutch.

The power unit kept the wire taut at all times so that the spacing between the two cars was measured by the position of the reel at each instant. This measurement was made by using a multiple-turn potentiometer geared to a reel shaft. A direct current generator tachometer operating the same shaft

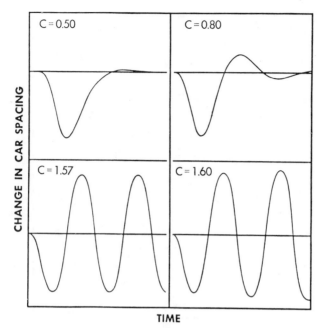

FIGURE 33. Change in car spacing of two cars with different values of $C = aT$ for the follower.

gave a measure of the rate at which the wire was wound or unwound, which was proportional to the relative speed of the cars. A fifth wheel, attached to the follower car, measured $u_f(t)$, while an accelerometer in the car indicated the longitudinal acceleration a_f.

All of this information was recorded simultaneously by an oscilloscope installed in the follower car, as shown in Figure 36.

The data were fitted to Eq. 27, and Table 9 shows results for eight drivers on a test track.

TABLE 9

CAR-FOLLOWING PARAMETERS

Driver	T (r = max), sec	a, sec^{-1}	r	$2aT$
1	1.4	0.74	0.87	2.08
2	1.0	0.44	0.90	0.88
3	1.5	0.34	0.86	1.03
4	1.5	0.32	0.49	0.97
5	1.7	0.38	0.74	1.29
6	1.1	0.17	0.86	0.37
7	2.2	0.32	0.82	1.43
8	2.0	0.23	0.85	0.93
Average	1.55	0.37		1.12

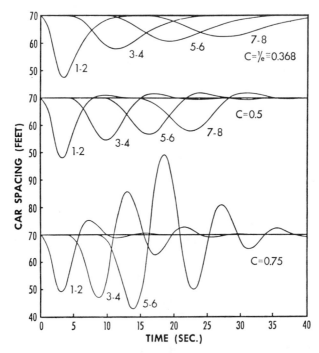

FIGURE 34. Car spacing of a line of cars with constant sensitivity, (a) for various values of T, and $C = aT$.

The lag time T was taken at the value given by the maximum correlation between the two sides of Eq. 27, and the value of the sensitivity coefficient a was computed for this lag. It can be seen that the correlation between the two sides of the equation were generally in the range 0.80–0.90.

The last column, $2aT$, is shown to indicate whether the asymptotic stability criterion $2aT < 1$ was met by the various drivers. The average of all drivers was close to the limit 1.12 versus 1.0. This agrees with the feelings of many drivers that high-speed, freeway-type driving is often close to the limit of stability. A few conservative drivers interspersed in a line of cars can add greatly to the stability of the stream because they leave large gaps that can absorb and greatly dampen disturbances.

Acceleration Noise

Drivers in a line of cars are influenced by the disturbances (bumps, curves, lapses of attention, and so on) that are the sources of the noisy behavior of the lead driver. The acceleration of the nth car can be expressed as the sum of two terms, one the car-following behavior and the other a random term

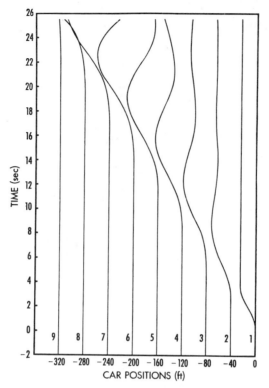

FIGURE 35. Illustrating asymptotic instability in a
line of nine cars in a moving coordinate system.

$b_n(t)$, which is the natural acceleration noise of the nth driver. This noise is
defined as the root mean square deviation of the acceleration of the car driven
independently of other vehicles.

Runs made on a section of the General Motors test track (an almost
perfect roadbed) by four operators, driving in the range 20–60 miles/hr
yielded Gaussian acceleration noise distributions with dispersions of
0.010 g $\pm$ 0.002 g, which are about 0.32 ft/sec^2.

It is quite clear that for a given driver the acceleration noise will vary
considerably as he drives on different roads or under different conditions.
The dispersion of the acceleration noise observed in a run in the Holland
Tunnel in New York was 0.73 ft/sec^2. Runs on roads with poor surfaces and
winding curves gave dispersions of 1.5–2 ft/sec^2.

Jones and Potts[76] devised a simple and ingenious method of measuring
acceleration noise making use of a tachograph, Kienzle type TCO8F. On the
tachograph, which records speed versus time in a continuous curve on a

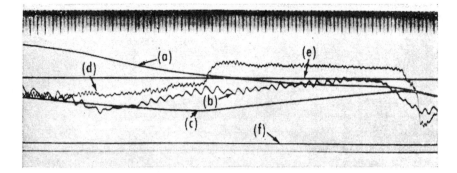

FIGURE 36. Reproduction of oscillograph recording for the car-following experiment. Curve (a) represents the car spacing, (b) the relative speed (c) the speed of the following car and, (d) the acceleration of the following car. The straight line (e) is the zero reference for the relative speed and acceleration of the following car, and (f) the reference for the car spacing and speed of the following car.

circular chart, the acceleration variation σ shows up as a measure of the smoothness of the speed–time curve. Because the smoothness of such a curve is a measure of the quality of the roadway on which it is made, σ becomes a measure of the poorness of a roadway. A narrow roadway with substandard curves and grades will force drivers to accelerate and brake frequently. The value of σ for such a road will be higher than that for a well-designed expressway.

Mathematically, σ may be defined as follows for an observation or trip time of duration T, starting at a time $t = 0$:

$$\sigma^2 = \frac{1}{T} \int_0^T [a(t) - a_{av}] \; dt,$$

where $a(t)$ is the cars acceleration at time t, and a_{av} is the average acceleration over time T. If the cars initial and final speed are the same, $a_{av} = 0$, and

$$\sigma^2 = \frac{1}{T} \int_0^T a(t) \; dt.$$

In making use of a tachograph, it is not easy to determine $a(t)$ directly and continuously. However, an approximation may be used in which

$$\sigma^2 = \frac{1}{T} \sum \left(\frac{\Delta v}{\Delta t}\right)^2 \Delta t$$

or

$$\sigma^2 = \frac{(\Delta v)^2}{T} \sum \frac{1}{\Delta t},$$

where Δt is the time taken to change Δv in speed, Δv being taken constant throughout the measurement of the record.

Jones and Potts obtained results by this method ranging 0.79–1.63 ft/sec², 0.0025–0.051 g.

Nonlinear Car-Following Models

The linear car-following model has considerable advantage in its mathematical simplicity, which permits the analytical solution to the stability question, as was previously described. However, when examined in terms of its implications for steady-state flow, particularly relative to the shape of the flow-concentration curve that would result from this law, it is not entirely satisfactory.

For purposes of investigating the steady-state implication of the model, one may neglect the time lags and write the expression in the following simplified form:

$$\frac{du}{dt} = a\frac{ds}{dt}, \qquad (28)$$

which says the acceleration of the average vehicle, du/dt, is proportional to the average relative speed ds/dt. This expression may be integrated to yield

$$u = as + c_1,$$

where c_1 is the constant of integration whose value is $c_1 = -as_j$, for $u = 0$, when traffic is stopped with a jam spacing of s_j; thus

$$u = a(s - s_j) = a\left(\frac{1}{k} - \frac{1}{k_j}\right),$$

$$q = ku = a\left(1 - \frac{k}{k_j}\right),$$

and the q–k curve is a straight line with a slope of $-a/k_j$. The intercept on the q axis is the maximum value of q, thus $a \rightarrow q_m$, when $u \rightarrow \infty$. This result does not agree as well with most observations of traffic flow as the Greenberg model derived from the fluid analogy, such as that shown in Figure 26, which shows a convex curve, not a straight line.

Gazis et al.[71] showed that the Greenberg steady-state model could be derived from car-following theory by a nonlinear model that included the inverse car spacing as well as the relative speed. The car-following equation is

$$\ddot{x}_{n+1}(t + T) = a \frac{[\dot{x}_n(t) - \dot{x}_{n+1}(t)]}{[x_n(t) - x_{n+1}(t)]},$$

which in the simplified form of Eq. 28 becomes

$$\frac{du}{dt} = \left(\frac{a}{s}\right)\frac{ds}{dt}. \tag{29}$$

Equation 29 can be integrated to yield

$$u = a \ln s + c_2, \tag{30}$$

where $s = s_j$, $u = 0$; thus $c_2 = -a \ln s_j$, and Eq. 30 may be written

$$u = a \ln \left(\frac{s}{s_j}\right) = a \ln \left(\frac{k_j}{k}\right),$$

which is Greenberg's model, and $a = u_m$, the speed at maximum flow.

This steady-state model also has a disadvantage in that the speed increases without limit as k approaches zero. This disadvantage and experimental results like those shown in Figure 27, led Edie[77] to propose a slightly more complicated model, which appeared to fit low concentrations better than the inverse spacing model. The model proposed by Edie, in the simplified form omitting time lags, is

$$\frac{du}{dt} = \left(\frac{au}{s^2}\right)\frac{ds}{dt}.$$

When this is integrated, and the coefficient a and the constant of integration are evaluated, it leads to the steady-state model:

$$k = k_m \ln \left(\frac{u_f}{u}\right).$$

A disadvantage of the Edie model is that the concentration increases without limit as the speed approaches zero. Therefore, it would not be expected to apply to very concentrated traffic. Also, for extremely light traffic the large spacings between vehicles would tend to make the car-following behavior less and less of a factor in determining traffic states. However, the model fits reasonably well within the range of concentration of 20 vehicles/mile up to k_m. The hypothesis that the steady-state behavior was different for concentrations less than optimum versus more than optimum was supported by studies of traffic on the Eisenhower Expressway in Chicago.[78]

The three car-following models mentioned thus far can be generalized by the introduction of two parameters, an exponent m on the speed of the $(n + 1)$st car and an exponent l on the spacing between the nth and $(n + 1)$st car (see Ref. 73). The general form is

$$\ddot{x}_{n+1}(t + T) = a\dot{x}_{n+1}^{m} \frac{[\dot{x}_n(t) - \dot{x}_{n+1}(t)]}{[x_n(t) - x_{n+1}]^{l}}. \tag{31}$$

The values of the exponents are the following for the three models:

Model	m	l
Linear	0	0
Inverse spacing	0	1
Speed, inverse spacing squared	1	2

Various other values might be chosen for these parameters. Herman and co-workers investigated values of m of 0, 1, and -1, and of l of $\frac{1}{2}$, 1, 2, 3, and $-\frac{1}{2}$.

For data taken by car-following runs in the Lincoln Tunnel in New York, Gazis et al.[73] found the best correlation for values of $m = 1$, $l = 2$ from among the three models above.

May and Keller[78] investigated fractional exponents for freeway and tunnel data and found those giving the best correlations were $m = 0.8$, $l = 2.8$ and $m = 0.6$, $l = 2.1$, respectively.

A generalization of the linear car-following model was made by Lee[90] by the addition of a memory function. Lee used an integral transform technique that makes it mathematically possible to consider a memory function defining the way in which a following driver processes his information. The assumption of the linear car-following model is retained, that is, the relative speed is the only factor influencing the follower car's acceleration. What is added is that the response depends on the time history of the relative speed and not just what it was a certain instant earlier. This is expressed mathematically by

$$\ddot{x}_n(t) = \int_0^t M(t - t')[\dot{x}_{n-1}(t') - \dot{x}_n(t')] \, dt'.$$

The memory function M contains two parameters λ' and T' that roughly corresponded to the sensitivity coefficient and time lag in the original linear model:

$$\lambda' = \int_0^\infty M(t) \, dt,$$

$$T' = \frac{1}{\lambda'} \int_0^\infty t M(t) \, dt.$$

Several functions may be considered for M, as follows:

$$M(t) = \lambda(t - T) \quad \text{Dirac delta.}$$

Physically this kind of a model implies that the following driver acts instantaneously to correct any relative velocity difference. It shows that a follower goes through a change in maneuver every interval of time nT, where n is an integer:

$$M(t) = ake^{-kt} \quad \text{decaying exponential.}$$

If a and k are positive constants, the roots of the characteristic equation always have negative real parts, implying that the motion is always locally stable and is asymptotically stable when $\lambda'T' < \frac{1}{2}$:

$$M(t) = ak^2te^{-kt}$$

is locally unstable when $\lambda'T' > 2$, and is asymptotically stable when $\lambda'T' < \frac{1}{4}$.

Rothery et al.[80] investigated the applicability of car-following models to buses. Using pairs of buses, car-following models, which had been shown to be valid for automobiles, were examined for buses and were found to provide a good representation of the detailed manner in which one bus follows another. Research on bus behavior is of importance in giving consideration to the establishment of exclusive bus lanes on freeways and parkways as a means of improving mass public transportation in urban areas.

The only difference in the experimental equipment used for the bus-following experiments was that of recording the data on magnetic tape, instead of oscillagraph paper, to be later converted into digital form for analyses.

A correlation study was carried out for the linear model ($l = m = 0$), the reciprocal spacing model ($l = 1, m = 0$) and the reciprocal spacing speed model ($l = 2, m = 1$). It was found that the correlation coefficient was generally quite high for all 22 test drivers for all three models, ranging from about 0.7 to 0.9. The response times T were low compared to car drivers, ranging from about 0.5 to 1.2 sec, with a mean about 0.8 sec, compared with a range of 1.0–2.2 sec, mean of 1.55 sec, for car drivers, as shown in Table 9. Sensitivity coefficients were about the same for buses, and therefore the stability criterion was far better met. The value of $C = aT$ was generally less than $1/e$, 0.368, giving nonoscillatory damped spacing responses.

Other Car-Following Models

While the relative speed models discussed above have been the most widely investigated, there are some other models which should be mentioned. In one version it is postulated that the $(n + 1)$st driver chooses a speed $u_{n+1}(t + T)$ at a time $t + T$ as some function of the spacing, that is,

$$u_{n+1}(t + T) = G_n[x_n(t) - x_{n+1}(t)] \tag{32}$$

for some suitable function G_n.

Equation 32 has the advantage of mathematical simplicity. The function G_n is chosen to give the observed steady-state relation between speed and spacing. There is an implication here as in previous models that the behavior of drivers under time-dependent conditions is determined by their behavior under conditions of steady flow.

Komentani and Sasaki[81] investigated stability problems with Eq. 32, where $G_n = a[x_n(t) - x_{n+1}(t)]$, with essentially the same results as obtained by Herman and co-workers.

Helly[82] studied a model of the form

$$\ddot{x}_{n+1}(t + T) = c_1[(\dot{x}_n - \dot{x}_{n+1}) + c_2(x_n - x_{n+1} - D)$$

$$+ c_3 B_n + c_4 B_{n-1}]_t, \tag{33}$$

where D is the desired spacing, and B_n and B_{n-1} are 1 or 0 depending on whether those cars are braking or not.

Helly used Eq. 33 to simulate the behavior of bottlenecks in a tunnel with mixtures of cars and trucks.

Rockwell et al.[83] investigated a piecewise linear curve fitting to a regression function:

$$\ddot{x}_{n+1}(t + T) = \ddot{x}_{nm} + b_0[s(t) - s_{nm}] + b_1[x_n(t) - \dot{x}_{nm}]$$

$$- b_2[\dot{x}_{n+1}(t) - \dot{x}_{(n+1)m}],$$

where $\ddot{x}_{nm}$, s_{nm}, $\dot{x}_{nm}$, and $\dot{x}_{(n+1)m}$ are the average value of acceleration, spacing, and speed for the nth car, and the speed for the $(n + 1)$st car, respectively. The averages were taken for the entire period of observation. The b_i are regression coefficients. By cutting the data into successively smaller subspaces, and postulating a linear model for the subspace, the effects of nonlinearity could be observed. It was found, however, that the effects of nonlinearity were not very important: the cutting procedure did not result in any substantial decrease of residual variance. The coefficient b_0 is usually small, positive, and independent of $s(t)$. The coefficients b_1 and b_2 are positive and independent of $s(t)$.

Application of Microscopic Theory

One of the important applications of car-following theory, other than studies of roadway safety and capacity, is the application to automated highways where the car-following process will be controlled by equipment instead of a human driver.

Bender and Fenton,[91] using car-following models, studied the flow capacity of highways. They determined that the following performance restrictions must be imposed on the steady-state car-following mode of an automatic vehicle control system.

1. The average separation between adjacent vehicles must not be excessive.
2. A controlled vehicle must be stable relative to the car ahead.
3. Asymptotic stability must be obtained.
4. The control system must not exceed the response capabilities of the vehicle.
5. Acceleration and deceleration must not exceed a limiting value of, say, 0.1 g.

A controlled vehicle's response to small changes in the lead-car speed can be described in terms of a small-signal vehicle time constant T. This parameter, which can be obtained from the system function equation, is a measure of how fast the controlled vehicle can respond to a small-signal input. The small-signal time constant is defined as the time required for the vehicle to reach 0.632 of the final incremented velocity value.

The change in spacing from a small-signal speed change v_s is

$$s = hv_s,$$

where h is a positive constant.

In order to avoid small spacings, the steady-state spacing S_0 is taken proportional to the absolute traffic speed V, that is,

$$S_0 = hV$$

and the new spacing, after a speed change, is

$$S = h(V + v_s).$$

The constant h is then the mean time headway.

The flow on the highway is

$$q = \frac{5280}{L/V + 1.465h},$$

where L is the car length, and the maximum flow is

$$q_m = \frac{5280}{L/V + 1.465 \times 0.787T}.$$

High flow rates cannot be obtained without a very responsive vehicle ($T \leq 1$), which would probably cause passenger discomfort. The vehicle time constant imposes a definite upper limit on the capacity of an automated highway.

Safety in an automated system has been studied and it has been found that the capacity and safety performance of a wayside control system was distinctly superior to that of a vehicle-borne system. For a desired lane capacity, the most suitable operating speed depends on vehicle length.

Car-Following and Accidents

Brill[84] recently investigated the effect of lower reaction times on the frequency of accidents, in particular, on rear-end collisions, using a car-following model involving driver reaction time, following distance, and deceleration response. He has related these parameters to the frequency of accidents.

The model, like most car-following models, assumes that a driver reacts only to the car directly in front, which tends to overestimate decelerations, but this is offset by assuming the choice of a minimum rate to avoid a collision. A principal result is

$$a_n = a_0 v_0 (v_0 - 2a_0 S_n)^{-1},$$

where

$$S_n = \sum_1^n (r_i - h_i),$$

and

$a_n =$ deceleration of nth car c_n,

$a_0 =$ deceleration of leading car c_0,

$v_0 =$ steady-state speed of all cars before c_0 deceleration,

$r_i =$ reaction time of ith car, $i = 1, 2, \ldots, n$,

$h_i =$ temporal headway of ith car.

Thus the deceleration rate chosen by c_n depends on his reaction times and temporal headways and the $n - 1$ preceding cars, through the sum S_n.

The probability of a collision depends on the distribution of the $(r_i - h_i)$, the maximum usable deceleration, and the value of a_0.

An interesting result is that an increase in mean value of $(r_i - h_i)$ results in a multiplicative change in collision probability. For example, if $v_0 = 50$ ft/sec, $A = 20$ ft/sec^2, $a_0 = 15$ ft/sec^2, $\bar{r} = 0.45$ sec, $\bar{h} = 0.65$ sec, $\sigma_r = 0.15$ sec, and $\sigma_h = 0.15$ sec, the probability of a collision is 0.0155 for some car to collide with the car ahead. If $\bar{r}$ increases to 0.55 sec, the probability of a collision increases by a factor of 8 to about 0.125. A is the maximum acceleration.

Weiss[79] has shown that there are difficulties in deducing a spacing distribution from car-following models, and one cannot hope to generate results anywhere as general for the car-following situation, as one can for the low-density (non–car-following) case, which tends toward a negative exponential distribution.

Microscopic traffic flow theories, describing in quantitative terms how one vehicle follows one or more other vehicles, offer considerable promise in dealing with traffic flow problems, accidents, and the eventual automation of the driving process. While current theories are oversimplified, they go a long way in predicting how one vehicle will respond to fluctuations in the speed of a vehicle ahead.

It seems likely that evolution of this theory will entail further complications, which will include more parameters describing the characteristics of the driver–vehicle–road complex.

VI. RECENT WORK

The amount of research dealing with flow on roads has reached a plateau in recent years. It is interesting that there were only three papers presented in this subject area at the Fifth International Symposium on the Theory of Traffic Flow and Transportation, held in June, 1971.[85] These were out of a total of 30 papers. This contrasts with 9 papers out of 15 at the first symposium held in December, 1959.

One of the recent papers was an experimental paper dealing with the starting characteristics of a platoon of vehicles.[86] There was a wealth of empirical results derived from observation of a 6-vehicle platoon. The lead driver in the platoon used combinations of three accelerations and three cruising speeds in a series of experiments starting from a standstill, accelerating to cruising speed, and coming to a stop after a distance of 0.7 mile.

Among the findings are the following: The starting delay decreased with higher acceleration. The speed of propagation of the starting speed wave with respect to the roadway increased from around 20–30 ft/sec backwards for speeds from 0 to around 20–30 ft/sec, then increased almost linearly to 25 ft/sec in the forward direction at the top speed of 75 ft/sec. A stationary wave occurred at 50 ft/sec, which, according to hydrodynamic theory, would be the speed at maximum flow. When the cruising speed was 40 or 50 miles/hr, the minimum transit time (maximum flow) for the platoon was around $8\frac{1}{2}$ sec. For cruising speeds of 30 miles/hr, the transit time dropped to around $7\frac{1}{2}$ sec, except for low acceleration.

The information reported on demonstrates a number of features that car-following and hydrodynamic models have not taken into account. This is important with respect to simulation calculations of traffic dynamics. Similar work was previously reported for platoon dynamics for cars and buses.[87,88]

Another recent paper modified car-following theory by adding a memory function. Most theories, as described herein, relate the acceleration of a following vehicle at a time t to some function of speeds, relative speeds, and distances at one earlier time, say $t - T$. In this paper,[6] it was assumed that a driver uses a weighted average of conditions at all previous times. The weighting function was found to have a sharp peak between 1 and 2 sec.

The third paper[89] provided new information regarding the effect of sample size on empirical relationships between traffic flow variables. It settled on the statistical result that if two variables are related linearly and both are subject to error, then the gradient of the least-squares regression line decreases as scatter increases. In this case the variables considered were average speed and traffic flow. There were striking differences in the slope for 6-sec, 30-sec, and 5-min time slices of the data.

There are three principal reasons for the reduced emphasis on flow theories in 1971 compared to 1959. They are, first, that most of the approaches taken to date have been explored mathematically with numerous variations, and therefore it is more difficult to contribute originally and significantly to the body of knowledge so far developed. Second, there is a growing emphasis on the application of knowledge to traffic engineering and management problems, reducing the gap between theory and practice. And third, the original, rather narrow, field of interest of transportation scientists has greatly broadened.

But the problem of developing a truly unified theory to describe the flow of traffic on roads still remains.

References

1. *Highway Capacity Manual*, Highway Research Board Special Report No. 87, Washington, D.C., 1965.
2. D. C. Gazis, R. Herman, and R. B. Potts, "Car-Following Theory of Steady-State Traffic Flow," *Operations Res.*, **7**, 499–505 (1959).
3. W. F. Adams, "Road Traffic Considered as a Random Series," *J. Inst. Civil Engrs.*, **4**, 121–130 (1936).
4. I. Prigogine, "A Boltzmann-like Approach to the Statistical Theory of Traffic Flow," in *Theory of Traffic Flow*, edited by R. Herman, Amsterdam: Elsevier, 1961.
5. J. G. Wardrop, "Some Theoretical Aspects of Road Traffic Research," *Proc. Inst. Civil Engrs.*, Part II, **1**, No. 2, 325–362 (1952).

6. J. N. Darrock and R. W. Rothery, "Car Following and Spectral Analysis," in *Traffic Flow and Transportation*, edited by G. F. Newell, New York: Elsevier, 1972.

7. E. C. Molina, *Poisson Exponential Bi-nomial Limit*, New York: Van Nostrand, 1942.

8. R. M. Oliver, "Distribution of Gaps and Blocks in a Traffic Stream," *Operations Res.*, **10**, 197–217 (1962).

9. G. F. Newell, "Stochastic Properties of Peak Short-Time Traffic Counts," *Transport. Sci.*, **2**, No. 3, 167–183 (1968).

10. M. C. Dunne, R. W. Rothery, and R. B. Potts, "A Discrete Markov Model of Vehicular Traffic," *Transport. Sci.*, **2**, No. 3, 233–251 (1968).

11. J. H. Kell, "A Theory of Traffic Flow on Urban Streets," *Proc. 13th Annual Western Section Meeting*, Inst. of Traffic Engrs., 66–70 (1969).

12. R. Rothery, lecture notes, "Theory of Traffic Flow," summer course, Purdue University, 1967.

13. R. F. Davison and L. A. Chimini, "The Hyperlang Probability Distribution— A Generalized Traffic Headway Model," University of Connecticut Civil Engineering Department, Report JHR 67-13.

14. D. J. Buckley, "Inter-Vehicle Spacing and Counting Distributions," Doctoral thesis, University of New South Wales, Australia, 1965.

15. D. J. Buckley, "Road Traffic Headway Distributions," *Proc. Aust. Road Res. Board*, **1**, 153–187 (1962).

16. A. Daou, "On Flow in Platoons," *Bull. Operations Res. Soc. Amer.*, 25th National Meeting, **12**, B40 (1964).

17. L. Breiman, A. V. Gafarian, R. Lichtenstein, and V. K. Murthy, "An Experimental Analysis of Single-Lane Time Headways in Freely Flowing Traffic," System Development Corporation Report, SD6 TN-3858/003/01, 1968, pp. 1–4.

18. D. J. Buckley, "A Semi-Poisson Model of Traffic Flow," *Transport. Sci.*, **2**, 107–133 (1968).

19. R. T. Underwood, "Traffic Flow and Bunching," *J. Aust. Rd. Res.*, **1**, 8–25 (1963).

20. A. J. Miller, "A Queueing Model for Road Traffic Flow," *J. Roy. Statist. Soc.*, **B-23**, 64–75 (1961).

21. A. J. Miller, "Traffic Flow Treated as a Stochastic Process," in Ref. 4, pp. 165–174.

22. A. J. Miller, "Analysis of Bunching in Rural Two-Lane Traffic," *Operations Res.*, **11**, 236–247 (1963).

23. J. Pahl and T. Sands, "Vehicle Interaction Criteria from Time Series Measurements," *Transport. Sci.*, **5**, 403–417 (1971).

24. B. Greenshields and A. Loutzenheiser, "Speed Distributions," *Proc. Highway Res. Board*, **20** (1940).

25. D. S. Berry and D. M. Belmont, "Distribution of Vehicle Speeds and Travel Times," in *Proc. 2nd Berkeley Symposium of Mathematical Statistics and Probability*, San Francisco: University of California Press, 1951.

26. J. Lindner, "A Contribution to the Statistical Analysis of Speed Distributions," in *Vehicular Traffic Science*, edited by L. Edie, R. Rothery, and R. Herman, New York: Elsevier, 1967, pp. 109–117.

27. M. T. Grace and R. B. Potts, "The Diffusion of Traffic Platoons," *Operations Res.*, **12**, No. 2, 255–275 (1964).
28. F. A. Haight, *Mathematical Theories of Traffic Flow*, New York: Academic, 1963, pp. 73–76, 114, 201, 215.
29. F. A. Haight and W. W. Mosher, "A Practical Method for Improving the Accuracy of Vehicular Speed Distributions," *Highway Res. Board Bull.*, 341 (1962).
30. I. Prigogine, R. Herman, and R. Anderson, "Further Developments in the Boltzmann-like Theory of Traffic," in *Proc. 2nd Intern. Symp. on the Theory of Road Traffic Flow*, edited by J. Almond, Paris: Office of Economic Co-operation and Development, 129–138 (1965).
31. I. Prigogine and R. Herman, *Kinetic Theory of Vehicular Traffic*, New York: Elsevier, 1971.
32. J. L. Kraus, "Speed Distributions," Port of New York Authority, R. & D. Report, 1966.
33. A. V. Gafarian, P. K. Munjal, and J. Pahl, "An Experimentive Validation of Two Boltzmann-Type Statistical Models for Multilane Flow," *Transport. Sci.*, **5**, 211 (1971).
34. R. Herman and T. Lam, "On the Mean Speed in the 'Boltzmann-Like' Traffic Theory: Analytical Derivation," *Transport. Sci.*, **5**, 314–327 (1971).
35. R. Herman and T. Lam, "On the Mean Speed in the 'Boltzman-Like' Traffic Theory: A Numerical Method", *Transport. Sci.*, **5**, 418–429 (1972).
36. R. Herman, T. Lam, and I. Prigogine, "Kinetic Theory of Vehicular Traffic-Comparison with Data," *Transport. Sci.*, **6**, 440 (1972).
37. G. Weiss and R. Herman, "Statistical Properties of Low-Density Traffic" *Quart. Appl. Math.*, **20**, 121–130 (1962).
38. G. F. Newell, "Mathematical Models of Freely Flowing Highway Traffic," *Operations Res.*, **3**, 176–186 (1955).
39. M. S. Bartlett, "Some Problems Associated With Random Velocities," *Publ. l'inst. Statistique de l'universite de Paris*, **6**, 261–270 (1957).
40. L. Carleson, "A Mathematical Model for Highway Traffic," *Nordisk Mat. Tidskr.*, **5**, 176–180 (1957).
41. T. W. Forbes, "Speed, Headway and Volume Relationships on A Freeway," *Proc. Inst. Traffic Engrs.*, 103–126 (1951).
42. J. C. Tanner, "An Improved Model for Delays on a Two-Lane Road," *J. Roy. Statist. Soc.*, **B-23**, 38–63 (1961).
43. E. Komentani, "On the Theoretical Solution of Highway Traffic Under Mixed Traffic," Jour. Faculty, Kyoto University, **17**, 79–88 (1955).
44. D. C. Gazis, R. Herman, and G. Weiss, "Density Oscillations Between Lanes of a Multi-Lane Highway," *Operations Res.*, **10**, 658–667 (1962).
45. Homer J. Holland, "A Stochastic Model for Multi-Lane Traffic Flow," *Transport. Sci.*, **1**, No. 3, 184–205 (1967).
46. M. J. Lighthill and G. B. Whitham, "On Kinematic Waves, II. A Theory of Traffic Flow on Long Crowded Roads," *Proc. Roy. Soc.* (*London*) **229A**, 317–345 (1955).

47. P. I. Richards, "Shock Waves on a Highway," *Operations Res.*, **4**, 42–51 (1956).
48. L. C. Edie and Eric Baverez, "Generation and Propagation of Stop–Start Traffic Waves," in Ref. 26, pp. 26–37.
49. L. C. Edie, "Discussion of Traffic Stream Measurements and Definitions," Ref. 30, pp. 139–154.
50. W. Leutzbach, "Testing the Applicability of the Theory of Continuity on Traffic Flow at Bottlenecks," in Ref. 26, pp. 1–13.
51. G. F. Newell, Instability in Dense Highway Traffic," in Ref. 30, pp. 73–83.
52. T. W. Forbes, M. J. Zagorski, E. L. Holshouser, and W. A. Deterline, Measurement of Driver Reactions to Tunnel Conditions," *Proc. Highway Res. Board*, Vol. **37**, 345–57 (1958).
53. R. Herman and R. B. Potts, Single-Lane Traffic Theory and Experiment," in Ref. 4, pp. 120–146.
54. H. Greenberg, An Analysis of Traffic Flow," *Operations Res.*, **7**, No. 1, 79–85 (1959).
55. L. C. Edie and R. S. Foote, Traffic Flow in Tunnels," *Proc. Highway Res. Board*, **37**, 334–344 (1958).
56. L. C. Edie and R. S. Foote, 'Experiments in Single-Lane Flow in Tunnels," in Ref. 4, pp. 175–192.
57. L. C. Edie and R. S. Foote, "Effect of Shock Waves on Tunnel Traffic Flow," *Proc. Highway Res. Board*, **39**, (1960).
58. R. S. Foote, "Single Lane Traffic Flow Control," in Ref. 30, pp. 84–103.
59. H. Greenberg and A. Daou, "The Control of Traffic Flow to Increase the Flow," *Operations Res.*, **8**, No. 4, 524–532 (1960).
60. J. G. Wardrop, "Experimental Speed/Flow Relations in a Single Lane," in Ref. 4, pp. 104–119.
61. R. E. Franklin, "Single-Lane Traffic Flow on Circular and Straight Tracks," in Ref. 26, pp. 42–55.
62. T. W. Forbes and M. E. Simpson, "Driver and Vehicle Responses in Freeway Deceleration Waves," *Transport. Sci.*, **2**, No. 1, 77–104 (1968).
63. J. Foster, "An Investigation of the Hydrodynamic Model for Traffic Flow with Particular Reference to the Effect on Various Speed–Density Relationships," *Proc. Aust. Road Res. Board*, **1**, 229–257 (1962).
64. R. J. Smeed, "Some Circumstances in Which Vehicles Will Reach Their Destinations Earlier by Starting Later," *Transport. Sci.*, **1**, No. 4, 308–317 (1967).
65. R. Reuschel, "Fahrzeugbewegungen in der Kolonne bei gleichformig beschleunigtem oder verzogertem Leitfarzeug," *Zeit. Osterr. Ing. und Arch. Ver.*, **95**, 52–62, 73–77 (1950).
66. L. A. Pipes, "An Operational Analysis of Traffic Dynamics," *J. Appl. Phys.*, **24**, 274–281 (1953).
67. E. W. Montroll and R. B. Potts, "Car-Following and Acceleration Noise," *An Introduction to Traffic Flow Theory*, Highway Research Board Special Report 79, 1964.
68. R. E. Chandler, R. Herman, and E. W. Montroll, "Traffic Dynamics: Studies in Car Following," *Operations Res.*, **6**, 165–184 (1958).

69. R. Rothery, R. Silver, and R. Herman, "Single-Lane Bus Flow," *Operations Res.*, **12**, 913–933 (1964).
70. R. Herman, W. Montroll, R. B. Potts, and R. W. Rothery, "Traffic Dynamics: Analysis of Stability in Car Following," *Operations Res.*, **7**, 86–106 (1959).
71. D. C. Gazis, R. Herman, and R. B. Potts, "Car-Following Theory of Steady-State Traffic Flow," *Operations Res.*, **7**, 499–505 (1959).
72. R. Herman and R. B. Potts, "Single-Lane Traffic Theory and Experiment," in Ref. 4, pp. 120–146.
73. D. C. Gazis, R. Herman, and R. W. Rothery, "Nonlinear Follow-the-Leader Models of Traffic Flow," *Operations Res.*, **9**, 545–567 (1961).
74. D. C. Gazis, R. Herman, and R. W. Rothery, "Analytical Methods in Transportation-Mathematical Car-Following Theory of Traffic Flow," *Proc. Amer. Soc. Civil Engrs., Mech. Div.*, **6**, 29–46 (1963).
75. R. Herman and R. W. Rothery, "Microscopic and Macroscopic Aspects of Single Lane Traffic Flow," *J. Operations Res. Soc. Japan*, **5**, 74–93 (1962).
76. T. R. Jones and R. B. Potts, "The Measurement of Acceleration Noise—a Traffic Parameter," *Operations Res.*, **10**, No. 6, 745–763 (1962).
77. L. C. Edie, "Car-Following and Steady-State Theory for Non-Congested Traffic," *Operations Res.*, **9**, No. 1, 66–76 (1961).
78. A. D. May and H. E. M. Keller, "Evaluation of Single and Multi-Regime Traffic Flow Models," in *Beiträge zur Theorie des Verkehrs flusses*, Bonn: Heransgegeben vom Bundesminister fur Verkekr, 1969, pp. 37–47; [Engl. trans.: Proc. 4th Intern. Symp. on Traffic Theory.]
79. G. Weiss, "On the Statistics of the Linear Car Following Model," *Transport. Sci.*, **3**, 88 (1969).
80. R. Rothery, R. Silver, and R. Herman, "Analysis of Experiments on Single-Lane Bus Flow," *Operations Res.*, **12**, No. 6, 913–933 (1964).
81. E. Komentani and T. Sasaki, "On the Stability of Traffic Flow," *J. Operations Res. Japan*, **2**, 11–26 (1958).
82. W. Helly, "Simulation of Bottlenecks," in Ref. 4, pp. 207–238.
83. T. H. Rockwell, R. L. Ernst, and A. Hanken, "A Sensitivity Analyses of Expirically Derived Car-Following Models," *Transportation Res.*, **2**, 363–373 (1968).
84. E. A. Brill, "A Car-Following Model Relating Reaction Times and Temporal Headways to Accident Frequency," *Transport. Sci.*, **6**, 343–353 (1972).
85. G. F. Newell, ed., *Traffic Flow and Transportation*, New York: Elsevier, 1972.
86. R. Herman, T. Lam, and R. Rothery, "Starting Characteristics of Automobile Platoons," in Ref. 85, pp. 1–17.
87. R. Herman, T. Lam, and R. Rothery, "Further Studies on Single Lane Bus Flow: Transient Characteristics," *Transport. Sci.*, **4**, 187–216 (1970).
88. Z. A. Nemeth and R. L. Vecellio, "Investigation of the Dynamics of Platoon Dispersion," *Highway Res. Board Rec.*, **334**, 23–33 (1970).
89. C. C. Wright, "Some Properties of the Fundamental Relations of Traffic Flow," in Ref. 85, pp. 19–32.
90. G. Lee, "A Generalization of Linear Car-Following Theory," *Operations Res.*, **14**, 595 (1966).

91. J. G. Bender and Robert E. Fenton, "On Vehicle Longitudinal Dynamics," in Ref. 85, pp. 19–32.
92. G. F. Newell, unpublished lecture notes, 1966.
93. R. M. Oliver, "A Two-Lane Traffic Model," Univ. of Calif. Bulletin ORC-64-34, 1–25, 1965.
94. Yasuji Makigami, G. F. Newell and R. H. Rothery, "Three-Dimensional Representation of Traffic Flow," *Transport. Sci.*, **5**, 303–313 (1971).

CHAPTER 2
Delay Problems for Isolated Intersections

Donald R. McNeil

and

George H. Weiss

++

Contents

I. Introduction   110

II. Characterization of Traffic Flow   110

III. The Gap Acceptance Function   114

IV. The Single-Car Delay Problem   117

V. The Pedestrian Queueing Problem   127

VI. The Vehicular Queueing Problem   134

VII. The Delay at a Traffic Signal   142

     The Arrival Process   143
     The Departure Process and Signal Discipline   144
     The Delay   145
     The Mean Delay for Poisson Input and Constant
       Departure Times   146
     The Mean Delay for a More General Arrival Process   148
     The Mean Delay for Random Departure Times   150
     The Overflow at the Traffic Signal   151
     The Optimization Problem   159
     The Vehicle-Actuated Traffic Signal   162
     Garwood's Model   169

References   169

I. INTRODUCTION

The delay in any automobile journey can be roughly described as the sum of the delay due to driver interaction with other moving vehicles and the delay due to fixed, regulatory devices typified by stop signs and traffic signals. Many models have been proposed to describe both kinds of delay, and, inevitably, all of these models fall short in important respects as descriptions of the many psychological and mechanical processes that determine the journey time. Nevertheless, the study of simplified models is valuable as an aid to data collection and organization of experiments and as a rough indication of the order of magnitude of changes in observed phenomena as functions of changes in relevant parameters. Such information is of direct importance to the traffic engineer who must perforce extrapolate from usually small numbers of data to a variety of anticipated situations.

It is the purpose of this chapter to review some—but obviously not all—of the research on delays at fixed signals. We further restrict our consideration to single signals rather than systems of such signals. Very little is known about the precision of measurements of individual driver characteristics required for reasonably accurate specification of a traffic system, but there has been considerable work on the interaction of drivers with individual signals. Hopefully, such work should be the cornerstone of a more comprehensive theory of traffic systems in much the same way as the study of the hydrogen atom and molecule has led to the understanding of far more complicated molecular systems.

Two problems, together with peripheral but related problems, will be discussed in this article. The first is the delay to single drivers and to vehicular queues at a stop or yield sign. The second problem is that of delay at a fixed traffic light. Most of the work on both of these problems is based on the theory of stochastic processes and, in particular, on the more specialized areas of renewal processes and queueing theory. For the reader interested in a good account of these topics slanted toward applications, few better books can be recommended than the monographs by Cox and Smith[1] and by Cox.[2]

II. CHARACTERIZATION OF TRAFFIC FLOW

It is possible to describe traffic along a highway in two ways: as a function of space and as a function of time. Both of these descriptions are useful in some circumstances, but in the present set of problems, where there is a fixed signal, it is obviously more useful to describe the flow of traffic in terms of the sequence of arrival times of successive cars on the road. Later, when we discuss the delay problem in greater detail, alternative descriptions will suggest themselves and will be discussed.

Let the sequence of arrival times of successive cars in an unimpeded stream at a fixed point in space be $t_1, t_2, t_3, \ldots$. The headways $\{G_r\}$ are defined by $G_r = t_r - t_{r-1}, r = 2, 3, 4, \ldots$, and G_1 is defined to be equal to t_1 even when $t = 0$ does not coincide with the arrival of a car on the main road. The assumption made in most traffic studies is that successive headways are independent and identically distributed random variables. In what follows we denote the common probability density function for $G_2, G_3, G_4, \ldots$ by $\varphi(G)$. If the instant $t = 0$ at which measurements are begun is chosen at random, that is, uncorrelated with the arrival of any car, then the probability density function for G_1, to be denoted by $\varphi_0(G)$, is given by[2]

$$\varphi_0(G) = \frac{1}{\mu} \int_G^\infty \varphi(x) \, dx, \tag{1}$$

where μ is the mean headway

$$\mu = \int_0^\infty G\varphi(G) \, dG. \tag{2}$$

In any meaningful model for traffic μ must be finite.

The most widely used form for $\varphi(G)$ in traffic studies is the negative exponential

$$\varphi(G) = \varphi_0(G) = \frac{1}{\mu} \exp\left(-\frac{G}{\mu}\right). \tag{3}$$

This density function was first observed experimentally by Adams,[3] and subsequent studies have shown that there is a theoretical justification for the use of Eq. 3 when the flow of traffic is light.[4-6] The proof given by Weiss and Herman* is the simplest, although Breiman and Thedeen prove their results under somewhat weaker restrictions. We therefore indicate the idea of the proof given by the former authors. Consider a stream of traffic on an infinitely long homogeneous highway. Suppose that each car is represented by a dimensionless point traveling at a constant speed chosen from a probability density function $f(v)$ which is not a delta function or a combination of delta functions (that is to say, not all cars travel at the same speed). Let us assume that when a car traveling at speed v overtakes another traveling at speed $v' < v$, it is able to pass without suffering a delay. Let us further assume that at $t = 0$ the cars are located on the highway in such a way that the probability that any car is located between x and $x + dx$ (measured in space) along the highway is $\rho_0(x) \, dx$, where $\lim_{x \to \pm\infty} \rho_0(x) = \rho_0 = $ constant. This would be true, for example, if successive headways were identically distributed random

* At the time this paper was being refereed, a manuscript by A. J. Miller with the same essential results was sent to one of us (GHW), indicating that the results were also found independently by Miller.

variables with a finite second moment. Since the dynamics of this model can be handled exactly, we can calculate the probability that at time t, the distance between a given car and any other car will be between x and $x + dx$. Call this function $\rho(x, t)\, dx$. A detailed consideration of the space and time relations between the two cars shows that

$$\rho(x, t) = \int_0^\infty \int_0^\infty \rho_0(x + (v - v')t) f(v) f(v')\, dv\, dv', \tag{4}$$

where v is the speed of the leading car and v' is the speed of the following car. In the limit $t \to \infty$ this becomes

$$\lim_{t \to \infty} \rho(x, t) = \rho_0 \int_0^\infty \int_0^\infty f(v) f(v')\, dv\, dv' = \rho_0. \tag{5}$$

Thus $\rho(x, \infty)$ is a constant. When this result is translated into the time domain, the probability density for the passage of two successive cars past a fixed point is

$$\varphi(t) = q \exp(-qt), \tag{6}$$

where q is the mean flow given by $q = \rho_0 \int_0^\infty v f(v)\, dv$.

Any extension of this result through the analysis of presumed driver and vehicular interactions leads to apparently intractable mathematical problems. It is difficult to state exactly the range of flow rates in which the flow might be described by a negative exponential form for $\varphi(t)$ as in Eq. 6, but an order of magnitude estimate of the upper limit for this figure is 1200 vehicles/hr/lane on a freeway with at least two lanes. Many empirical headway distributions have been proposed and analyzed in the literature. All of them make some attempt to deal with the problem of very small gaps, which are assigned a relatively high weight by the negative exponential distribution, but which are in reality zero because of car size. The simplest of these models was first proposed by Schuhl,[7] who suggested the function

$$\varphi(t) = 0 \qquad\qquad\qquad t \le T,$$
$$= q \exp[-q(t - T)] \qquad t > T. \tag{7}$$

Many more elaborate distributions have been proposed, in particular, by Buckley.[8,9] None of these have any theoretical base, making it difficult to predict their range of applicability. It is obvious, however, that the more parameters allowed into a hypothesized distribution, the more likely it will be to fit observed distributions.

The properties of successive headways can be described in ways other than through the assumption that the headways are independent random variables. Miller,[10] for example, developed the notion of traveling queues. His work starts from the observations that the effect of traffic interaction is often to bunch cars together on parts of the highway, leaving gaps between successive bunches or queues. The definition of a queue is arbitrary, but for some data provided by the Swedish State Roads Institute, Miller defined a queue by requiring that the (time) headway between successive cars be no more than 8 sec and that the relative speed of cars within a queue lie within the range -3 to 6 miles/hr. Successive queues were assumed to be independent, and the spacing between queues was assumed to follow a negative exponential distribution. Good agreement between this model and the observed data was verified in detail. The distribution of the number of cars in each queue was fitted about equally well by the distributions

$$p_n = \frac{n^{n-1}}{n!} r^{n-1} e^{-rn},$$

$$p_n = \frac{(m+1)(m+1)!(n-1)!}{(m+n-1)!}, \tag{8}$$

where r and m are parameters fitted from the data. One can derive an expression for p_n under the assumptions of successive independent headways. If a queue is defined as being a succession of cars no more than T sec apart, and if we define the tail probabilities $\Phi(T)$ by

$$\Phi(T) = \int_T^\infty \varphi(t)\, dt, \tag{9}$$

then p_n is given by

$$p_n = [1 - \Phi(T)]^{n-1} \Phi(T). \tag{10}$$

It is interesting that the second expression for p_n given in Eq. (8) can be derived from this distribution by assuming that T is a suitably distributed random variable.[11]

Another characterization of traffic sometimes found in the literature is in terms of blocks and gaps.[12] The terminology dealing with this definition varies between authors, but the basic idea is that a gap in a stream of traffic is the period of time between two successive vehicles whose headway is greater than a time T, while a block is the period of time between two successive cars whose headway is less than T. This characterization of traffic flow is useful in discussing the intersection delay problem in the situation where a waiting driver in a minor stream of traffic will merge if the headway is greater than T but not if it is less than T. The paper by Oliver[12] discusses several statistics related to the block-gap description of traffic. There is no difficulty in extending the present description by assuming that a headway t is a gap with probability $\alpha(t)$ and a block $1 - \alpha(t)$. Indeed, this is done implicitly in analyses of the vehicular queueing problem. Finally we mention the process of binomial arrivals, which is more of mathematical than practical interest.[13]

In this process time is treated as consisting of discrete intervals of duration h. The parameter h is chosen to be sufficiently small that no more than one arrival per interval can occur. The probability of an arrival in any given interval is p, and this event is independent of events in other intervals. Hence the number of arrivals during interval j is N, which is a random variable with a discrete binomial distribution. Thus, the expectation of the number of vehicles arriving in k disjoint intervals is kp, and the variance is $kp(1 - p)$. The binomial distribution can be regarded as a discrete analog of the negative exponential distribution, since in the latter, the probability of a single arrival in an infinitesimal time interval $(t, t + dt)$ is $\lambda \, dt$, with λ the arrival rate. This is independent of events at other times, and no more than one arrival is permitted in the interval dt.

III. THE GAP ACCEPTANCE FUNCTION

In order to develop the theory of merging and queueing problems related to delays at stop and yield signs we are obliged to discuss the factors which govern driver behavior in the minor stream. The simplest problem, that of the delay to a single driver who wishes to merge with a single stream of traffic at a stop sign requires the specification of a rule that describes the circumstances under which the waiting driver is willing to merge with the traffic. It is generally assumed with some experimental evidence,[14-16] that the decision is based solely on the size of the headway, measured in units of time, perceived by the minor road driver. If the gap is denoted by G, then the information is summarized in terms of a gap acceptance function $\alpha(G)$. This function gives the probability of a merge with the given gap. Earliest studies of the intersection delay problem made the assumption that $\alpha(G)$ could be represented as a step function

$$\alpha(G) = H(G - T), \tag{11}$$

where $H(x) = 0$ for $x < 0$, and $H(x) = 1$ for $x > 0$.[3,17,18] That is to say, merging takes place only during gaps greater than T.

Measurements of $\alpha(G)$, of course, reveal a much more complicated situation. The simplest "laboratory" measurements of $\alpha(G)$ were made by Herman and Weiss,[19] who found that $\alpha(G)$ could be represented by

$$\alpha(G) = 0 \qquad\qquad G < T,$$
$$= 1 - \exp\left[-\lambda(G - T)\right] \qquad\qquad G \geq T, \tag{12}$$

where, for the conditions of the experiment, the parameters λ and T were found to be $\lambda = 2.7 \text{ sec}^{-1}$ and $T = 3.3$ sec. The experiments that led to these results were performed under highly idealized conditions in which none of the many interacting factors that determine the gap acceptance function in real traffic was present. It is therefore difficult to infer quantitative information about real populations from such limited experiments. There have been a number of measurements of gap acceptance functions under circumstances not subject to experimental control. The situation is somewhat complicated in nonlaboratory experiments since one is forced to measure a distribution of gap acceptance functions and cannot repeat measurements on any driver. Theoretical questions related to the estimation of gap acceptance function from field measurements have been discussed by McNeil and Morgan.[20] All of the field measurements are made on population gap acceptance functions $\alpha_p(G)$, under the assumption that the $\alpha(G)$ for a single driver is a step function $H(G - T)$, where T is a random variable. Thus, if $u(T)$ is the probability density function of T, the measured $\alpha_p(G)$ is

$$\alpha_p(G) = \int_0^\infty H(G - T)u(T)\,dT = \int_0^G u(T)\,dT. \qquad (13)$$

This restriction on the individual gap acceptance function has the slight disadvantage that it predicts that no driver will ever accept a gap smaller than one previously rejected. Extensive measurements of the gap acceptance function for populations of drivers have been published by Blunden, Clissold, and Fisher.[21] They reported good agreement between data and an $\alpha_p(G)$ of the form

$$\alpha_p(G) = \frac{1}{\Gamma(n + 1)} \int_0^{G/G_0} x^n e^{-x}\,dx, \qquad (14)$$

where G_0 was estimated to be 3.1 and 3.5 sec for two locations at which the values of n were found to be 5.6 and 7.2, respectively.

Possibly the most extensive set of measurements of gap acceptance functions has been made by Drew and his collaborators[22,23] and Wattleworth et al.[24] at the Texas Transportation Institute. They were primarily interested in merges made onto a freeway, generally from an input ramp. Since cars do not necessarily come to a full stop when entering freeway traffic, a distinction needs to be drawn between merges made from a full stop and moving merges. The engineering terminology used for drawing a distinction between these is to call a full headway (a headway between two cars in the major traffic stream) a *gap* and the time between the arrival of a car at the merge point on the feeder road and the arrival of the first car in the main traffic stream, a *lag*. Drew and his co-workers[22,23] presented several more definitions relevant to the types of merge that are possible from an acceleration lane. The data are presented under the assumption that population (as opposed to individual) gap acceptance functions can be represented in terms of a log–normal function. This type of data reduction was suggested in a study of lag and gap acceptances at stop-controlled intersections by Solberg and Oppenlander.[25] They showed that the probit* of the percent of the population accepting a time gap G is related to G by

$$Y = a + b \log G, \tag{15}$$

in which a and b are constants. Extensive tables of $\alpha(G)$ expressed in this form are presented for both single-vehicle merges and for multiple merges of a queue of vehicles. The data were gathered in several American cities. Wattleworth et al.[24] also presented data on parameters characterizing ramp design and on the statistics of accepted or rejected gaps.

It is probable that in the most general situation the gap acceptance function depends on parameters other than just the gap. Wagner[26] has shown experimentally that $\alpha_p(G)$ can depend on the amount of traffic, the type of gap (i.e., gap or lag), and vehicle type. Worrall et al.[27] performed a discriminant analysis of several possible factors influencing the gap acceptance function. They found that out of nine factors considered, only two had a significant effect. These were ramp vehicle speed and the gap in the main traffic stream. Cohen, Dearnaley, and Hansel[28] studied population gap acceptance functions for pedestrians, fitting their results to the log–normal form.

Recently Miller[29] presented an enlightening comparative study of nine different estimators of some parameters of the gap acceptance function. Specifically, Miller assumed a population of drivers, each having a gap acceptance function of step function form, but with a distribution of critical gaps. Using artificial data, Miller tried to estimate the mean critical gap by each of nine suggested methods. He found that two methods of the nine had a reasonably small bias, a method described by Ashworth[30,31] and a maximum likelihood method developed by Miller himself. The principal finding is that the maximum likelihood method gives the best results, but depends on an assumed distribution. Miller did test for robustness against a change in distribution form that assumed (changing a log–normal distribution to a normal) and found only a small change in the estimated mean. Ashworth's technique proved only slightly less accurate and leads to simpler computations, which can be done on a desk calculator.

Two areas in the elucidation of gap acceptance functions remain largely unexplored. The first is the problem of determining the form of $\alpha(G_1, G_2, \ldots, G_n)$ for a driver wishing to cross several lanes of traffic. The second is a thorough investigation of robustness properties of models of delay, making use of various forms of gap acceptance function. It is clear that $\alpha(G)$ has no intrinsic interest, but must be related to questions that are

---

* The probit Y of a probability p is defined to be the solution to $p = (2\pi)^{-1/2} \int_{-\infty}^{Y} \times \exp(-u^2/2) \, du$.

of interest to the city planner and designer of traffic control equipment. There is some evidence to suggest that very simple models which use step gap acceptance functions are sufficiently accurate for engineering purposes.[32-39] If this is found to be true in general, very elaborate measurements of $\alpha_p(G)$ would not greatly contribute to useful engineering information.

IV. THE SINGLE-CAR DELAY PROBLEM

Having now discussed the various functions characterizing drivers and traffic, we are in a position to calculate the statistics of delay to a single driver who arrives at a stop sign at $t = 0$ and is delayed by one or more lanes of traffic on an intersecting main road. In the single-lane merge problem we will assume that successive headways in the major stream of traffic are identically distributed independent random variables with probability density $\varphi(t)$. The driver on the side road will be assumed to arrive at the intersection at $t = 0$, and his gap acceptance function will be denoted by $\alpha(t)$. We will calculate an expression for $\Omega(t)$, the probability density of delay on the feeder road. The work to be described follows closely analysis presented by Weiss and Maradudin.[40] Earlier analyses of special cases of the problem as just stated were given by Adams,[3] Raff,[17] Tanner,[18] and Mayne.[41]

The delay time can either be zero, with an acceptable gap being present at $t = 0$, or it can be positive, which implies that several cars on the main road have passed the stop sign, and a merging maneuver takes place immediately after the passage of a car on the main roads. The probability of an initial acceptable gap will be denoted by $\bar{\alpha}_0$ and is

$$\bar{\alpha}_0 = \int_0^\infty \alpha(t)\varphi_0(t) \, dt. \tag{16}$$

Hence we can express $\Omega(t)$ in the form

$$\Omega(t) = \bar{\alpha}_0 \delta(t) + \Omega_1(t), \tag{17}$$

where $\delta(t)$ is a Dirac delta function and $\Omega_1(t)$ represents the contribution due to a delayed merge. The probability that an arbitrary gap is acceptable is $\bar{\alpha}$, where $\bar{\alpha} = \int_0^\infty \alpha(t)\varphi(t) \, dt$, so that

$$\Omega_1(t) = \bar{\alpha}w(t), \tag{18}$$

in which $w(t) \, dt$ is the probability that a car on the main road passes the intersection in the time interval $(t, t + dt)$ conditional on no merge having been made. It will now be shown that $w(t)$ is the solution to an integral equation of convolution type. For the derivation let us define two functions

$$\Psi_0(t) = \varphi_0(t)[1 - \alpha(t)],$$
$$\Psi(t) = \varphi(t)[1 - \alpha(t)]; \tag{19}$$

$\Psi'_0(t) \, dt$ is the probability that the first gap is between t and $t + dt$, and is unacceptable, and $\Psi'(t) \, dt$ has the same interpretation for succeeding gaps. If a car in the main stream passes the intersection at t and no merge has been made, it is either the first to do so, or the last such event occurred at some $\tau < t$ and the succeeding gap was unacceptable. These two possibilities lead to the equation

$$w(t) = \Psi'_0(t) + \int_0^t w(\tau)\Psi'(t - \tau) \, d\tau. \tag{20}$$

Since this equation contains a convolution integral, the use of Laplace transforms is suggested. Let us denote the Laplace transform of a function of t by that same function of s with an asterisk, so that, for example,

$$\Omega^*(s) = \int_0^\infty e^{-st}\Omega(t) \, dt,$$
$$\Psi'^*(s) = \int_0^\infty e^{-st}\Psi'(t) \, dt. \tag{21}$$

The Laplace transform of Eq. 20 leads to

$$w^*(s) = \frac{\Psi_0'^*(s)}{1 - \Psi'^*(s)} \tag{22}$$

so that the combination of Eqs. 17, 18, and 22 implies

$$\Omega^*(s) = \bar{\alpha}_0 + \frac{\bar{\alpha}\Psi_0'^*(s)}{1 - \Psi'^*(s)}. \tag{23}$$

There are few cases of interest for which this transform can be inverted to yield an explicit expression for $\Omega(t)$, but the main information of interest lies in the moments of delay time. These are easily obtained from Eq. 23 by using it as a moment-generating function:

$$\overline{t^n} = \int_0^\infty t^n\Omega(t) \, dt = (-1)^n \frac{d^n}{ds^n} \Omega^*(s)\Big|_{s=0}. \tag{24}$$

Using this formula we find for the first two moments of the delay:

$$\bar{t} = \int_0^\infty t\left[\Psi'_0(t) + \left(\frac{1 - \bar{\alpha}}{\bar{\alpha}}\right)\Psi'(t)\right] dt,$$
$$\overline{t^2} = \int_0^\infty t^2\left[\Psi'_0(t) + \left(\frac{1 - \bar{\alpha}_0}{\bar{\alpha}}\right)\Psi'(t)\right] dt + \frac{2}{\bar{\alpha}}\bar{t}\int_0^\infty x\Psi'(x) \, dx, \tag{25}$$

The second of these expressions corrects an error in Eq. 17 of Weiss and Maradudin.[40]

When $\varphi(t) = (1/\mu) \exp(-t/\mu)$ and $\alpha(t)$ is the step function $\alpha(t) = H(t - T)$, the expression for $\Omega^*(s)$ is

$$\Omega^*(s) = e^{-T/\mu} \frac{s + (1/\mu)}{s + (1/\mu) \exp\{-[s + (1/\mu)]T\}}, \tag{26}$$

which, expanded and inverted term by term, leads to

$$\Omega(t) = e^{-t/\mu}\delta(t) + e^{-T/\mu} \sum_{n=1}^{\infty} \frac{(-1)^n}{(n-1)!\,\mu^n} \{e^{-nT/\mu}(t - nT)^{n-1}H(t - nT)$$

$$- e^{-(n-1)T/\mu}[t - (n-1)T]^{n-1}H[t - (n-1)T]\}, \tag{27}$$

which is not very convenient for use in applications. The mean and variance of delay time are easily found from $\Omega^*(s)$ to be

$$\bar{t} = \mu\left(e^{T/\mu} - 1 - \frac{T}{\mu}\right),$$

$$\sigma^2 = \mu^2\left[e^{2T/\mu} - \left(\frac{2T}{\mu}\right)e^{T/\mu} - 1\right]. \tag{28}$$

The preceding analysis is easily modified to deal with the case of a yield sign for which the driver tends to accept a shorter first gap than if he had come to a complete halt. For this situation let $\alpha_0(t)$ be the gap acceptance function for the first gap and $\alpha(t)$ be that for succeeding gaps. Then in Eqs. 20 and 25, $\bar{\alpha}_0$ and $\Psi_0(t)$ are to be replaced by

$$\bar{\alpha}_0 = \int_0^{\infty} \alpha_0(t)\varphi_0(t)\,dt,$$

$$\Psi_0(t) = \varphi_0(t)[1 - \alpha_0(t)]. \tag{29}$$

In the special case specified by

$$\alpha_0(t) = H(t - T_0),$$

$$\alpha(t) = H(t - T), \tag{30}$$

$$\varphi(t) = \frac{1}{\mu} \exp\left(-\frac{t}{\mu}\right)$$

the first and second moments of the delay time are

$$\bar{t} = \mu\left(e^{T/\mu} - e^{(T-T_0)/\mu} - \frac{T}{\mu} + \frac{(T - T_0)}{\mu} e^{-T_0/\mu}\right),$$

$$\sigma^2 = \mu^2\left[\left(e^{T/\mu} - \frac{T}{\mu}\right)^2(1 - e^{-2T_0/\mu}) - \frac{2T_0}{\mu}\left(e^{T/\mu} - \frac{T}{\mu}\right)e^{-2T_0/\mu}\right. \tag{31}$$

$$\left. - \frac{T_0^2}{\mu^2} e^{-T_0/\mu}(1 + e^{-T_0/\mu}) - \frac{T^2}{\mu^2} e^{-T/\mu}(1 - e^{-T_0/\mu})\right].$$

The derivation outlined here can also be applied to derive moments of the waiting time, for an impatient driver, characterized by an ensemble of gap acceptance functions

$$\alpha_1(t) \leq \alpha_2(t) \leq \alpha_3(t) \leq \cdots, \tag{32}$$

where $\alpha_j(t)$ is the gap acceptance function for the gap preceding the jth arrival on the main highway. Detailed expressions are given by Weiss and Maradudin.[40] Expressions and graphs of the expected delay time and the variance for the situation characterized by

$$\alpha(t) = H(t - T),$$

$$\varphi(t) = \frac{\sigma^{n+1}t^n}{n!} \exp(-\sigma t) \qquad n = 0, \ldots, 3, \tag{33}$$

are tabulated by Drew et al.[22,23] When the gap acceptance function is of the form

$$\alpha(t) = \{1 - \exp[-\lambda(t - T)]\}H(t - T) \tag{34}$$

the mean waiting time with a negative exponential headway distribution is

$$\frac{\bar{t}}{\mu} = e^{T/\mu} - 1 - \frac{T}{\mu} + \frac{1}{\lambda\mu}\left[e^{T/\mu} - 1 - \frac{T}{\mu} + \left(\frac{1}{1 + \lambda\mu}\right)^2\left(1 + \frac{T}{\mu} + \lambda T\right)\right]$$

$$\times (1 - e^{-T/\mu}) + \frac{e^{-T/\mu}}{1 + \lambda\mu} + \frac{T}{\mu}e^{-T/\mu}\bigg]. \tag{35}$$

Notice that this result reduces to that given in Eq. 28 in the limit $\lambda = \infty$. Some recent work by Blumenfeld and Weiss[39,40] shows that if one defines a critical gap T_c such that the expressions for $\bar{t}$ are equal for an underlying $\alpha(t)$ of the form given in Eq. 34 or a step gap acceptance function, then the variances are quite close to one another, and the two expressions for the probability of an immediate merge $\bar{\alpha}$ are also experimentally undistinguishable. This result holds for the negative exponential headway distribution and suggests that the theory developed with a step gap acceptance function is quite adequate for most engineering purposes.

Two other studies of the effects of different types of gap acceptance functions also tend to support the conclusion that the simple theory appears to be quite robust. The first examines the additional delay time caused by a driver who takes two consecutive gaps into consideration. More specifically, if he finds an acceptable gap he also considers the next following gap with some probability, and waits for that gap if it is larger.[36] Specific calculations were made for a step gap acceptance function, a negative exponential headway distribution, and for a probability of considering a second gap of the form

$$p(G) = \exp[-\rho(G - T)]H(G - T), \tag{36}$$

where T is the critical gap and ρ is a parameter which characterizes the probability. The calculations indicate that the expected added delay time is generally negligible in comparison to the expected wait for the first gap, the only possible exception being for $\rho = 0$ with very light traffic, so that gaps are very long. This, however, is an unrealistic situation for the model, so that the general conclusion of robustness can be drawn.

The second generalized gap acceptance function is that which depends both on the gap in time and the speed of the oncoming vehicle on the main road.[37] We therefore denote the gap acceptance function by $\alpha(v, t)$, and let $\varphi(v, t)\, dv\, dt$ be the joint probability for the headway between two successive cars to be between t and $t + dt$, and the velocity of the second car (relative to a fixed point) to be between v and $v + dv$. The theory of delay at a stop sign which has been outlined in earlier paragraphs can be adapted to the present case provided that we make the replacements

$$\varphi_0(v, t) = \frac{\displaystyle\int_t^\infty \varphi(v, \tau)\, d\tau}{\displaystyle\int_0^\infty dt \int_0^\infty t\varphi(v', t)\, dv'},$$

$$\bar{\alpha} = \int_0^\infty \int_0^\infty \varphi(v, t)\alpha(v, t)\, dv\, dt, \tag{37}$$

$$\Psi(t) = \int_0^\infty \varphi(v, t)[1 - \alpha(v, t)]\, dv,$$

with analogous substitutions for $\bar{\alpha}_0$ and $\Psi_0(t)$. Detailed calculations have been made for

$$\varphi(v, t) = \varphi_0(v, t) = \frac{f(v)}{\mu} \exp\left(-\frac{t}{\mu}\right), \tag{38}$$

that is, for the speed–headway distribution appropriate to light traffic. When $\alpha(v, t)$ can be expressed as

$$\alpha(v, t) = H(t - T(v)), \tag{39}$$

the mean delay is given by

$$\bar{t} = \frac{\mu}{\bar{\alpha}} \left\{ 1 - \int_0^\infty [1 + \lambda T(v)]f(v)e^{-T(v)/\mu}\, dv \right\}. \tag{40}$$

When $T(v)$ is linear in v,

$$T(v) = \frac{T_0 v}{\bar{v}}, \tag{41}$$

where $\bar{v}$ is the mean speed, and $f(v)$ is

$$f(v) = \frac{a^{r+1}v^r}{\Gamma(r+1)} e^{-av},$$
(42)

where, in terms of the mean and variance of $f(v)$,

$$a = \frac{\bar{v}}{\sigma^2},$$

$$r = \left(\frac{\bar{v}}{\sigma}\right)^2 - 1,$$
(43)

the mean delay is

$$\frac{\bar{t}}{\mu} = \frac{1}{\bar{\alpha}} - 1 - \frac{T_0\bar{v}^2}{\mu\bar{v}^2 + T_0\sigma^2}.$$
(44)

In this formula, $\bar{\alpha}$ is

$$\bar{\alpha} = \left(1 + \frac{T_0\sigma^2}{\mu\bar{v}^2}\right)^{-(\bar{v}/\sigma)^2}$$
(45)

The mean delay calculated from Eq. 40 for $T_0 = 5\,\text{sec}$ is presented for several typical parameters in Table 1. The first line, $\sigma = 0$, reproduces the result given in Eq. 28, and the second and third lines give the effects of a dispersion of velocities. Similar calculations have been made for $T(v) = T_0(v/\bar{v})^2$ and for other speed distributions. All of these calculations tend to support the contention that the speed dependence in the gap acceptance function does not introduce changes in the mean delay which are large enough to be measured. It is probable, too, that when traffic is heavy enough so that light traffic approximation inherent in the use of Eq. 38 is no longer valid, then variations in speed will not be as pronounced. Hence one might conjecture that in no case will the dependence of gap acceptance function on speed be important.

TABLE 1

MEAN DELAY IN SECONDS AS A FUNCTION OF FLOW RATE

		Flow Rate			
		600/hr	800/hr	1200/hr	1600/hr
$\sigma = 0$		1.8	4.3	7.9	13.1
$\bar{v} = 30$	$\sigma = 10$	1.8	4.1	6.9	10.4
$\bar{v} = 55$	$\sigma = 10$	1.8	4.2	7.6	12.3

McNeil and Smith[42] compared results on queueing delay obtained with two models: one in which each driver had a step gap acceptance function and another in which there was a distribution of gap acceptance functions. This comparison was also made in a paper by Blumenfeld and Weiss[38] on the robustness of various calculations of intersection delay to underlying assumptions. Ashworth[30] has also considered the effects of a distribution of critical gaps on the capacity of an intersection. Unfortunately, it is difficult to compare the results of these three studies because of slightly different assumptions in each. It appears that major differences will show up only at high traffic flows (that is, greater than 1200 cars/hr). Further experiments would indeed be valuable in the elucidation of the appropriateness of different models.

So far we have described the intersection delay problem in terms of the headways between successive cars. Miller[10] has suggested what might be called a "random queue" model for the same type of problem. This model makes the assumption that the major road traffic can be described as a succession of traveling queues of vehicles. The definition of a queue is somewhat arbitrary (see Section III for a brief discussion), but operationally we may say that no merge is possible while a queue is passing the merge point. We let the probability density function (p.d.f.) for the time of passage of a queue be $q(t)$, and let the p.d.f. of the headway (in this case defined to be the difference in time between the end of one queue and the beginning of the next queue) be $\varphi(t)$. Further, we define the mean duration of a queue to be $v [= \int_0^\infty tq(t)\,dt]$ and let μ be the mean associated with $\varphi(t)$. Then the probability of arrival during an interval between two successive queues is found from renewal theory[2] to be

$$p = \frac{\mu}{\mu + v}. \tag{46}$$

If we again let $\Omega(t)$ be the p.d.f. of delay time, and $\Omega^*(s)$ be its Laplace transform, then $\Omega^*(s)$ can be found by an argument similar to that leading to Eq. 23 to be

$$\Omega^*(s) = p\bar{\alpha}_0 + \frac{p\bar{\alpha}q^*(s)\Psi_0^*(s) + (1 - p)\bar{\alpha}q^*(s)}{1 - q^*(s)\Psi^*(s)}, \tag{47}$$

where $q^*(s)$ is the Laplace transform of $q(t)$. The mean delay is found from this expression to be

$$\bar{t} = \frac{(1 - \bar{\alpha}_0 p)v}{\bar{\alpha}} + p\int_0^\infty t\Psi_0(t)\,dt + \frac{1 - \bar{\alpha}_0 p}{\bar{\alpha}}\int_0^\infty t\Psi(t)\,dt, \tag{48}$$

and the probability of no delay is

$$p_0 = p\bar{\alpha}_0. \tag{49}$$

Equation 25 for the random car model is obtained from Eq. 48 by setting $v = 0$, which implies that $p = 1$. Miller compared the results of his random queue model with some data on pedestrian crossings in Sweden, finding that the mean delay was predicted slightly better by the random queues model than by the random cars model, and that the probability of immediate crossing was definitely better predicted by the random queues model. The model has been generalized by Weiss[43] to take into account mixed traffic.

Some of the earlier work on the delay problem is phrased in terms of gaps and blocks.[12,17,18] In this context a block is a period of time during which no merge will be made, and a gap is a period of time during which a waiting driver will choose to merge. Although this type of description of traffic flow has some mathematical advantages, it also requires a joint specification of the properties of traffic flow and driver characteristics. Hence it would be difficult to translate into experimental terms.

Up to this point we have discussed the merging problem for merges made into a single lane of traffic. Much less is known both experimentally and theoretically about the problem of crossing a multilane highway. The general form of the gap acceptance function is $\alpha(G_1, G_2, \ldots, G_n)$ for an n-lane highway,* but without further restrictions it appears to be impossible to solve the delay problem. The first theoretical results were due to Tanner,[18] who solved the problem for a negative exponential distribution of headways in each lane and a gap acceptance function of the form

$$\alpha(G_1, G_2, \ldots, G_n) = H(G^* - T), \tag{50}$$

where $G^* = \min_n(G_1, G_2, \ldots, G_n)$. However, measurements of two-lane gap acceptance functions indicate that α is *not* a symmetric function of the gaps.[21,26] In fact, it is plausible to assume asymmetry, since relatively larger gaps should be required in lanes further away from the waiting driver.

More recently another analysis of the multilane single-car crossing problem has appeared by Gazis et al.[44] The two principal assumptions contained in the theory are

1. The gap acceptance function can be written in factorized form as

$$\alpha(G_1, G_2, \ldots, G_n) = \alpha_1(G_1)\alpha_2(G_2) \cdots \alpha_n(G_n). \tag{51}$$

2. The headway distribution in each lane is negative exponential, the mean headway in lane j being denoted by μ_j.

---

* Again, we will presume that gap acceptance depends only on the gap measured in time and on no other parameters. This is much more of an assumption in the present case than it is in the single-lane merge since there have been no experimental data bearing on the matter.

The analysis of the model as presently stated is still quite difficult without the addition of another slightly more artificial assumption, namely, that the driver waiting at the merge point orders the given set of gaps so that $G_{j_1} \leq G_{j_2} \leq \cdots \leq G_{j_n}$, and the gaps are examined in sequence from shortest to longest. If G_{j_r} is the shortest unacceptable gap, the time to the next regeneration point of the underlying renewal process is taken to be G_{j_r} and no knowledge is assumed about $G_{j_{r+1}}, G_{j_{r+2}}, \ldots, G_{j_n}$.

As before, we let $\Omega(t)$ be the p.d.f. of waiting time and $\Omega^*(s)$ be its Laplace transform. The probability of zero delay is $\bar{\alpha}_1 \bar{\alpha}_2 \cdots \bar{\alpha}_n$, where $\bar{\alpha}_r$ is the probability that the gap in lane r is acceptable, that is,

$$\bar{\alpha}_r = \frac{1}{\mu_r} \int_0^\infty \alpha_r(x) \exp\left(-\frac{x}{\mu_r}\right) dx. \tag{52}$$

Suppose now that the waiting time is greater than zero. Then it is possible to define regeneration points as we have indicated above, as the duration of the smallest unacceptable gap. Let $\rho(t)$ be the p.d.f. for the time of arrival of the first unacceptable gap, and let $\rho^*(s)$ be its Laplace transform. Then we can write

$$\Omega^*(s) = \bar{\alpha}_1 \bar{\alpha}_2 \cdots \bar{\alpha}_n + \rho^*(s)\Omega^*(s) \tag{53}$$

or

$$\Omega^*(s) = \frac{\bar{\alpha}_1 \bar{\alpha}_2 \cdots \bar{\alpha}_n}{1 - \rho^*(s)}. \tag{54}$$

Now only the calculation of $\rho(t)$ is required. Suppose that the gap in lane r is unacceptable. Then either the first car to arrive in any other lane arrives after the one in lane r, or else it arrives before and is found to be acceptable. Hence we can write

$$\rho(t) = \sum_{r=1}^n \frac{1}{\mu_r} e^{-t/\mu_r}[1 - \alpha_r(t)] \prod_{\substack{k=1 \\ k \neq r}}^n \left(e^{-t/\mu_k} + \frac{1}{\mu_k}\int_0^t e^{-x/\mu_k}\alpha_k(x)\,dx\right). \tag{55}$$

It is not difficult to derive an expression for $\rho(t)$ when $\alpha_r(t) = H_r(t - T_r)$ and from there to find moments of the delay. If we let

$$\frac{1}{\nu_r} = \frac{1}{\mu_r} + \frac{1}{\mu_{r+1}} + \cdots + \frac{1}{\mu_n}, \tag{56}$$

then the expected delay is

$$\bar{t} = \exp\left(\frac{T_1}{\mu_1} + \frac{T_2}{\mu_2} + \cdots + \frac{T_n}{\mu_n}\right)$$
$$\times [\nu_1 + \exp(-T_1/\nu_1)(\nu_2 - \nu_1) + \exp[-(T_1/\mu_1 + T_2/\nu_2)](\nu_3 - \nu_2)$$
$$+ \exp[-(T_1/\mu_1 + T_2/\mu_2 + T_3/\nu_3)](\nu_4 - \nu_3) + \cdots$$
$$- \exp[-(T_1/\mu_1 + T_2/\mu_2 + \cdots + T_n/\mu_n)](\mu_n + T_n)]. \tag{57}$$

Fortunately, it is possible to obtain an approximation to this formula that appears to be quite accurate in practice. The approximation consists of using the formula for the delay for a single lane with $\alpha(t) = H(t - \bar{T})$ and $\varphi(t) = (1/\bar{\mu}) \exp(-t/\bar{\mu})$, where the parameters $\bar{\mu}$ and $\bar{T}$ are given by

$$\frac{1}{\bar{\mu}} = \frac{1}{\mu_1} + \frac{1}{\mu_2} + \cdots + \frac{1}{\mu_n},$$

$$\bar{T} = \bar{\mu}\left(\frac{T_1}{\mu_1} + \frac{T_2}{\mu_2} + \cdots + \frac{T_n}{\mu_n}\right). \tag{58}$$

A comparison of some results of evaluating the exact formula given in Eq. 57 and the indicated approximate formula is given in Table 2. As can be seen from the entries, the approximation gives results that are experimentally indistinguishable from the accurate ones. It remains to be seen whether this conclusion is true for a wider class of gap acceptance functions. It appears

TABLE 2

COMPARISON OF THEORETICAL AND APPROXIMATE CALCULATIONS OF MOMENTS OF DELAY FOR THE n LANE DELAY PROBLEM

No. of Lanes	μ_j, sec	T_j, sec	Theoretical $\bar{t}$, sec	$\overline{t^2}$	Approximate $\bar{t}$	$\overline{t^2}$
2	$\mu_1 = 8$ $\mu_2 = 8$	$T_1 = 5$ $T_2 = 7$	12.61	373.6	12.01	336.0
2	$\mu_1 = 4$ $\mu_2 = 6$	$T_1 = 5$ $T_2 = 9$	29.55	1854	28.54	1723
3	$\mu_1 = 12$ $\mu_2 = 8$ $\mu_3 = 8$	$T_1 = 5$ $T_2 = 7$ $T_3 = 8$	20.08	881.1	19.80	855.6
3	$\mu_1 = 8$ $\mu_2 = 6$ $\mu_3 = 6$	$T_1 = 5$ $T_2 = 7$ $T_3 = 8$	41.08	3512	40.66	3437
4	$\mu_1 = 12$ $\mu_2 = 8$ $\mu_3 = 8$ $\mu_4 = 8$	$T_1 = 5$ $T_2 = 7$ $T_3 = 8$ $T_4 = 8$	49.69	5108	49.30	5025
4	$\mu_1 = 8$ $\mu_2 = 6$ $\mu_3 = 6$ $\mu_4 = 6$	$T_1 = 5$ $T_2 = 7$ $T_3 = 8$ $T_4 = 8$	130.06	34,205	129.4	33,861

from the results of simulation that dropping the assumption that the minor stream driver orders the gaps in the major stream does not radically change the present results.

These theoretical calculations do not, of course, completely solve the n-lane delay problem, even with the assumption of a negative exponential distribution of headways in each lane. Certainly further experimental work is required to clarify the form of $\alpha(G_1, G_2, \ldots, G_n)$, and to examine the approximations made. Other forms of gap acceptance functions yield models amenable to analysis. For example, one can solve a model in which a gap acceptance function depends only on the gap formed by two successive arrivals and the two arrival lanes. Such a model leads to the consideration of semi-Markov models, which were mentioned by Weiss and Maradudin.[40] Whether such a model leads to very different results from those given above remains an open question, but we are inclined to doubt that such is the case. Another area of research that is relatively untouched involves nonstationary flow on the main road. An important application of the theory of delay by nonstationary traffic is that of the effects of traffic lights on the delay at intersections not directly controlled by the lights. A related question was treated by Garwood,[45] but the only other formulation of the problem suggests that a discouraging amount of numerical work would be required to fill this gap in our knowledge.[40] No completely satisfactory theory yet exists for the effects of an acceleration lane, although Haight, Bisbee, and Wojcick[46] have discussed some relevant points and Drew and co-workers[22,23] have collected data on merges from acceleration lanes. Recently, Mine and Mimura[47] have developed the theory of merges made from an infinitely long acceleration lane. Further results on this problem were obtained by Blumenfeld and Weiss.[48]

V. THE PEDESTRIAN QUEUEING PROBLEM

In the last section we discussed the ability of a stream of traffic to delay a single driver. If there is any amount of traffic on the main road, the delay to a driver can be decomposed into the delay while at the head of a queue (the service time) and the delay in queue. In heavy traffic the latter can easily be many times the former. As we shall see, parameters additional to those described in the last section may be required to specify the queueing model. Since little experimental information is available on highway traffic queues, the theories to be discussed are more speculative than those in the last section. Nevertheless, the presently available results are probably sufficient for many engineering purposes. The work to be described will be relevant only to the merging problem and not to delay by traffic lights, which will form the subject matter of later sections.

One of the first merge queueing problems to be discussed, and certainly the simplest of its genre, is the pedestrian queueing problem as analyzed by Tanner.[18] In this paper Tanner assumed a single lane of traffic* with a negative exponential headway distribution, a Poisson input of pedestrians at the street corner, and a common gap acceptance function for all of the pedestrians. Furthermore, the pedestrians could cross as a group without limitation on the size of the group. The reader will recognize that the waiting time of a single pedestrian is just that calculated in the last section, for a single car at a merge point. Tanner's contribution was in the calculation of the statistics of queue size and other parameters to be described below. Our discussion of the problem will follow a more recent analysis[35,49] that relaxes some of the restrictions used by Tanner. We will assume a single lane of traffic, a general stationary headway distribution, and an unrestricted gap acceptance function. It will be assumed that a pedestrian either crosses immediately on arrival at the intersection or else joins the waiting group.

The method of solution to be used is that of the embedded Markov chain. For regeneration points we choose the times of arrival of cars at the merge point. Transition probabilities p_{nm} are defined for the event "m people in queue at a regeneration point, given n people in queue at the preceding regeneration point." Two cases must be taken into account in the calculation of p_{nm}. When $n > m$ the transition $n \to m$ can take place only in the sequence $n \to 0 \to m$, that is, the group of n crosses as a whole; then at least m pedestrians join the queue, of which exactly m remain in queue. If $n \le m$ the transition $n \to m$ can occur either as $n \to 0 \to m$, or as $n \to m$, since the initial group does not necessarily cross. We can therefore decompose p_{nm} as

$$p_{nm} = p_{nm}^{(1)} + p_{nm}^{(2)}, \tag{59}$$

where $p_{nm}^{(1)}$ accounts for the indirect transition $n \to 0 \to m$ and $p_{nm}^{(2)}$ accounts for the direct transition $n \to m$. By our previous comment $p_{nm}^{(2)}$ is identically zero for $n > m$. Let us consider a headway of t. By our assumption of the Poisson arrival of pedestrians, the probability that an arriving pedestrian will see an initial gap between x and $x + dx$ is dx/t, that is, it is uniformly distributed over the entire interval. Therefore, the probability that a pedestrian arriving at some time in $(0, t)$ will cross immediately is $\alpha^*(t)$, where

$$\alpha^*(t) = \frac{1}{t} \int_0^t \alpha(x) \, dx. \tag{60}$$

We first calculate $p_{nm}^{(1)}$. Let the arrival rate for pedestrians be λ. If the gap is t, the waiting group crosses with probability $\alpha(t)$, and $k + m$ pedestrians

---

* Multiple lanes were treated by using the gap acceptance function $\alpha(G_1, G_2, \ldots, G_n) = H(G^* - T)$, where $G^* = \min_j (G_1, G_2, \ldots, G_n)$.

arrive with probability

$$\frac{e^{-\lambda t}(\lambda t)^{k+m}}{(k+m)!}$$

Of these, m remain in queue and k cross on arrival. These observations imply that $p_{nm}^{(1)}$ is given by

$$p_{nm}^{(1)} = \sum_{k=0}^{\infty} \int_0^{\infty} \alpha(t)\varphi(t) \frac{(\lambda t)^{k+m}}{(k+m)!} e^{-\lambda t} \binom{k+m}{m} [\alpha^*(t)]^k [1 - \alpha^*(t)]^m \, dt$$

$$= \frac{1}{m!} \sum_{k=0}^{\infty} \frac{1}{k!} \int_0^{\infty} \alpha(t)\varphi(t)[\lambda t \alpha^*(t)]^k [\lambda t(1 - \alpha^*(t))]^m \, dt$$

$$= \frac{1}{m!} \int_0^{\infty} \alpha(t)\varphi(t)[T(t)]^m e^{-\lambda T(t)} \, dt, \tag{61}$$

where

$$T(t) = \lambda t[1 - \alpha^*(t)] = \lambda \int_0^t [1 - \alpha(x)] \, dx. \tag{62}$$

When $m \geq n$, the transition $n \to m$ can take place in a manner described by Eq. 61, or the original group did not cross, and $k + m - n$ pedestrians arrived, of whom $m - n$ remain, in addition to the original group. This possibility yields an expression for $p_{nm}^{(2)}$ as

$$p_{nm}^{(2)} = \sum_{k=0}^{\infty} \binom{k+m-n}{m-n} \frac{1}{(k+m-n)!} \int_0^{\infty} [1 - \alpha(t)]\varphi(t)e^{-\lambda t}(\lambda t)^{k+m-n}$$

$$\times [\alpha^*(t)]^{k+m-n}[1 - \alpha^*(t)]^{m-n} \, dt$$

$$= \int_0^{\infty} [1 - \alpha(t)]\varphi(t) \frac{(\lambda T(t))^{m-n}}{(m-n)!} e^{-\lambda T(t)} \, dt. \tag{63}$$

If we define parameters Δ_n and ϵ_n by

$$\Delta_n = \frac{1}{n!} \int_0^{\infty} \alpha(t)\varphi(t)[\lambda T(t)]^n e^{-\lambda T(t)} \, dt,$$

$$\epsilon_n = \frac{1}{n!} \int_0^{\infty} [1 - \alpha(t)]\varphi(t)[\lambda T(t)]^n e^{-\lambda T(t)} \, dt, \tag{64}$$

then the transition matrix P can be written

$$P = \begin{bmatrix} \Delta_0 + \epsilon_0 & \Delta_1 + \epsilon_1 & \Delta_2 + \epsilon_2 & \cdots \\ \Delta_0 & \Delta_1 + \epsilon_0 & \Delta_2 + \epsilon_1 & \cdots \\ \Delta_0 & \Delta_1 & \Delta_2 + \epsilon_0 & \cdots \\ \Delta_0 & \Delta_1 & \Delta_2 & \cdots \\ \vdots & \vdots & \vdots & \end{bmatrix} \tag{65}$$

The equilibrium properties of the Markov chain defined by this transition matrix can be obtained quite easily. Let $\boldsymbol{\theta} = (\theta_0, \theta_1, \theta_2, \ldots)$ be the vector of steady-state probabilities; that is to say, $\boldsymbol{\theta}$ is the solution to

$$\boldsymbol{\theta} = \boldsymbol{\theta}P. \tag{66}$$

Writing this equation in component form we see that

$$\theta_n = \Delta_n + \sum_{r=0}^{n} \theta_r \epsilon_{n-r}, \tag{67}$$

which is easily solved by generating functions. If we define

$$\theta(z) = \sum_{n=0}^{\infty} \theta_n z^n,$$

$$\Delta(z) = \sum_{n=0}^{\infty} \Delta_n z^n = \int_0^{\infty} \alpha(t)\varphi(t)e^{-\lambda T(t)(1-z)}\, dt, \tag{68}$$

$$\epsilon(z) = \sum_{n=0}^{\infty} \epsilon_n z^n = \int_0^{\infty} [1 - \alpha(t)]\varphi(t)e^{-\lambda T(t)(1-z)}\, dt,$$

then Eq. 67 implies that

$$\theta(z) = \frac{\Delta(z)}{1 - \epsilon(z)}. \tag{69}$$

Moments of the queue length can be derived from these last two equations by differentiation. General expressions for the mean queue length and associated variance are

$$\bar{n} = \lambda\mu\left(\frac{1 - \bar{\alpha}_0}{\bar{\alpha}}\right),$$

$$\sigma^2 = \bar{n} + \frac{\lambda^2}{\bar{\alpha}}\int_0^{\infty} T^2(t)\varphi(t)\, dt - \frac{\lambda^2}{\bar{\alpha}^2}\left[\int_0^{\infty} T(t)\alpha(t)\varphi(t)\, dt\right]^2. \tag{70}$$

For $\alpha(t) = H(t - T)$ and $\varphi(t) = (1/\mu)\exp(-t/\mu)$, these formulas reduce to

$$\bar{n} = \lambda\mu(e^{T/\mu} - 1),$$

$$\sigma^2 = \lambda\mu(1 + 2\lambda\mu)e^{T/\mu} - \lambda\mu\left(\frac{1 + 2\lambda\mu + 2T + \lambda T^2}{\mu}\right) \tag{71}$$

as first given by Tanner.[18]

Similar results can be obtained for the statistics of queue length at a random time, that is, at a time uniformly distributed over a headway, rather than at a regeneration point. It is found that the mean number of pedestrians in queue at a random time is

$$\bar{n} = \lambda \bar{t} \tag{72}$$

using the general result of Little,[50] where $\bar{t}$ is given by the expression in Eq. 25. For a negative exponential headway distribution and a step gap distribution the expression for $\bar{n}$ is

$$\bar{n} = \lambda \mu \left(e^{T/\mu} - 1 - \frac{T}{\mu} \right) \tag{73}$$

as first derived by Tanner.[18]

There are a number of defects in the theory presented so far as a description of reality. One of these, the assumption that all pedestrians have the same gap acceptance function, can be remedied fairly easily. Let us suppose that the gap acceptance function depends on the gap and on a random variable v with associated probability density function $g(v)$, and let the rate parameter $\lambda(v)$ also depend on v. Then, for example, Eq. 72 should be changed to

$$\bar{n} = \int \lambda(v) \bar{t}(v) g(v) \, dv. \tag{74}$$

Unless very heavy traffic intensities are involved, the corrections due to a random variation of gap acceptance functions are negligible.[35] A further deficiency in the theory is the fact that the gap acceptance function is independent of group size. If one introduces a set of gap acceptance functions $\{\alpha_n(t)\}$, such that $\alpha_n(t)$ is the relevant function for a group of n, the problem cannot be solved in closed form. However, if it is assumed that only a finite number of the $\alpha_n(t)$ are different, then the solution is relatively straightforward. For example, if

$$\alpha_1(t) = \beta(t),$$
$$\alpha_2(t) = \alpha_3(t) = \cdots = \alpha(t), \tag{75}$$

where it is assumed that each arriving pedestrian makes a decision to cross on the basis of $\beta(t)$ but thereafter makes decisions on the basis of $\alpha(t)$, then

$$\theta(z) = \frac{\Delta(z) + \theta_1 \gamma(z)}{1 - \epsilon(z)}, \tag{76}$$

where

$$\gamma(z) = (1 - z) \int_0^\infty \varphi(t)[\beta(t) - \alpha(t)] \exp\left[-\lambda(1 - z)T_1(t)\right] dt,$$

$$T_1(t) = \int_0^t [1 - \beta(u)] \, du, \tag{77}$$

$$\theta_1 = \frac{\Delta_1(1 - \epsilon_0) + \Delta_0 \epsilon_1}{(1 - \epsilon_0)^2 - \gamma_1(1 - \epsilon_0) - \gamma_0 \epsilon_1}.$$

There is not enough experimental information available to warrant investigation of models any more elaborate at the present time.

Tanner[18] considered three alternative procedures for pedestrian crossing of two lanes of traffic, in an investigation of possible advantages of pedestrian islands. The three procedures are: (1) waiting for a gap of T in the combined traffic; (2) waiting for a gap of T in the near stream followed immediately by a gap of T' in the far stream; (3) waiting for a gap of T in the near stream, crossing to a pedestrian island, then waiting for a gap of T in the far stream. Of the first two methods considered, a small number of data suggested that method (1) was in fact used by pedestrians.

Smeed[51] has published some data on pedestrian crossing in which he related the percentage of pedestrians using pedestrian crossings to the traffic flow along an 1800-ft length of road, deducing a linear relationship. He has also presented data on the delay to vehicles at zebra crossings.* The dependent parameter studied was the journey time for a car over a stretch of road on either side of the crossing, and the independent parameter was the average number of pedestrians crossing per unit of time. A linear relation between these two parameters was found. If vehicular queueing effects are ignored and pedestrian flow is sufficiently small, a simple theoretical model predicts this relationship. Suppose that the interval over which measurements are made is large enough so that the travel time in motion over it is a constant t_r independent of whether a stop has been made or not. Then the total travel time is equal to $t_r + \bar{t}$, where $\bar{t}$ is the expected delay at the zebra crossing. Assuming that the arrival of pedestrians is described by a Poisson process with arrival rate λ and that the driver has a step gap acceptance function, we may use the expression given in Eq. 28 for $\bar{t}$ with $\mu = 1/\lambda$. If λT is small, we can expand the exponential to find the approximation $\bar{t} \sim \lambda T^2/2$, so that the total expected travel time is $t_r + \lambda T^2/2$, and is therefore proportional to λ as found by Smeed. The value of T suggested by his data lies between 5.4 and 6.5 sec. It is probably true that an analogous expression can be found for more general forms of $\alpha(t)$.

A recent paper by Thedeen[94] has considered the statistics of delay at pedestrian crossings where there are push-button controls. The specific model allows for a Poisson arrival of pedestrians at a corner at which there is an indicator which can register either red (stop) or green (go) for the pedestrian. A time t_b or greater must elapse between the end of one green period and the start of the succeeding one. When the control button is pushed, a period of at least t_a must elapse before the start of the next green period. This time will be exactly t_a provided the first pedestrian arrives after

---

* A zebra crossing is a pedestrian crossing across one or several streams of traffic in which the pedestrian has absolute priority. The name refers to the stripes painted in the streets in the British system.

a time $t_b - t_a$ (assumed positive) measured from the end of the last green period. If he arrives at time $t \leq t_b - t_a$, the next green period starts at t_b following the end of the last green period; that is, the pedestrian waits a time $t_b - t$ for the start of the next green period. It is also assumed that a pedestrian arriving during a green period will always be able to cross. Thedeen derives three results by methods that can be considerably simplified. These are the expected delay to a single pedestrian, the expected total pedestrian delay over a long period of time, and the expected number of pedestrians to cross during a green period. As an example of the mathematical techniques that can be used, we shall calculate the last of these. Let us suppose that a green period ends at $t = 0$. Then the first pedestrian to arrive does so either in the time interval $(0, t_b - t_a)$ or in $(t_b - t_a, \infty)$. In the first instance, if the first pedestrian arrival occurs at $t(\leq t_b - t_a)$, the time to the beginning of the next green period is $t_b - t$ and the probability that there are r arrivals in this period is $\lambda^r (t_b - t)^r \exp[-\lambda(t_b - t)]/r!$ If the first pedestrian arrival occurs at $t(\geq t_b - t_a)$, then the time to the beginning of the next green period is t_a, so that the probability of r arrivals in the period $(t, t + t_a)$ is $\lambda^r t_a^r e^{-\lambda t_a}/r!$. Hence the probability that r pedestrians arrive during the period when the light is red is

$$
\begin{aligned}
p_r &= \lambda \int_0^{t_b - t_a} e^{-\lambda t} \frac{\lambda^{r-1}(t_b - t)^{r-1}}{(r-1)!} e^{-\lambda(t_b - t)} dt \\
&\quad + \lambda \int_{t_b - t_a}^\infty e^{-\lambda t} \frac{(\lambda t_a)^{r-1}}{(r-1)!} e^{-\lambda t_a} dt \\
&= e^{-\lambda t_b} \left(\frac{\lambda^r t_b^r}{r!} - \frac{\lambda^r t_a^r}{r!} + \frac{\lambda^{r-1} t_a^{r-1}}{(r-1)!} \right) \qquad r = 1, 2, \ldots . \quad (78)
\end{aligned}
$$

Therefore the expected number of pedestrians who cross during the combination of a red and a green light is

$$
\bar{n} = \lambda(t_b + t_g) + \exp[-\lambda(t_b - t_a)]. \quad (79)
$$

Thedeen[94] has also shown that the expected waiting time of a single pedestrian is

$$
\bar{w} = \frac{t_a + \lambda(t_b^2/2) \exp[\lambda(t_b - t_a)]}{1 + \lambda(t_b + t_g) \exp[\lambda(t_b - t_a)]}, \quad (80)
$$

and the limiting value of the expected total waiting time measured over a long total period of time t is just $\lambda t \bar{w}$. Furthermore, the long-term proportion of time spent in the green cycle (i.e., the fraction of time during which cars are delayed) is

$$
p_g = \frac{\lambda t_g}{\lambda(t_b + t_g) + \exp[-\lambda(t_b - t_a)]} = \frac{n_g}{\bar{n}}, \quad (81)
$$

where $\bar{n}_g$ is the expected number of arrivals during the green period and $\bar{n}$ is the denominator of this last expression.

The most careful studies of the cost of pedestrian delays and consequent economic benefits of building zebra crossings or pedestrian subways or flyovers, have been made by Smeed.[51,52] It must be stated at the onset that the figures produced in any such study depend critically on the values assigned to both pedestrian and vehicle delay and will obviously vary both within the same country and from country to country. Furthermore, since the underlying empirical data on increased vehicle travel time due to pedestrians are for specific English locations,[52,53] the generalizations to other locations can only be made by taking analogous measurements at those locations. The formula given by Smeed for increased vehicle travel due to P pedestrians crossing at an uncontrolled crossing, with W being the width of the road and v the traffic speed (in miles/hr), is

$$D = \left(\frac{4.7 \times 10^{-2}}{v^2} - 8.9 \times 10^{-4}\right)PW. \tag{82}$$

This formula is valid for speeds of less than 20 miles/hr, that is, for heavy city traffic. In order to compare this with the cost of building pedestrian crossing facilities, it must be converted to monetary units. The suggested conversion factor for Great Britain is 12 shillings and 6 pence/vehicle hr ($1.50/vehicle hr). On the assumption of constant vehicle and pedestrian flows 10 hr/day and 300 days/yr, the annual loss in pounds to vehicle owners and occupants caused by λ pedestrians/hr is approximately

$$L = \left(\frac{87}{v^2} - 0.168\right)W\lambda. \tag{83}$$

If this expression is compared with the building cost of pedestrian crossing facilities, one can derive a criterion for the economic justification of building, let us say, a pedestrian subway. These calculations take no account of pedestrian delay or of possible accident prevention. Studies at the Road Research Laboratory indicate that the economic loss due to pedestrian delay or due to accidents, is usually negligible in comparison to the economic costs of vehicular delay.

VI. THE VEHICULAR QUEUEING PROBLEM

Problems of vehicular queueing present considerably more difficulty mathematically than the class of models just considered. However, the theoretical literature on the subject is fairly large, notwithstanding a paucity of supporting experimental data. The vehicular queueing problem can be categorized as a bulk queueing model in which servicing occurs at random times and for random periods, and in which the number of cars "serviced" by a given gap depends on the gap and the number of cars in queue. Although this description can also be applied to the pedestrian queueing model, a new feature must now be taken into account, the possibility that only a portion of any queue may be able to merge during a single gap. Thus, vehicular queues can be recurrent or transient. That is to say, it is possible for the number of cars in queue to build up indefinitely, whereas in our formulation of the pedestrian queue, the expected queue size remains bounded as a function of time because of the assumption that the entire group of pedestrians can cross the road in any acceptable gap. If it is assumed that the input on the feeder road is described by a Poisson process with parameter λ, then there exists a critical input λ_c such that when $\lambda > \lambda_c$ the queue increases without limit, and for $\lambda < \lambda_c$ the queue remains finite with probability 1.

It is reasonably simple to find a formal expression for λ_c.[32] For the calculation we require information about the time to move from the second position in queue to the first in queue and about the gap acceptance process. We will assume that in general the move-up time Δ will be a random variable with probability density $g(\Delta)$, and that the gap acceptance process will be as described for the single-vehicle merge problem. We need the further assumption that if the first driver in line merges during a gap G, the second driver in line sees a gap $G - \Delta > 0$; this is to say, the move-up time is always smaller than an acceptable gap. The critical rate λ_c can be written in terms of $M(t)$, the expected number of cars from an infinite queue that merge in time t as

$$\lambda_c = \lim_{t \to \infty} \frac{M(t)}{t}. \tag{84}$$

It is possible to show that this limit exists, and that λ_c can be calculated as the ratio of the expected number of cars to merge in a single gap to the expected headway.[32] If $\alpha_n(t)$ denotes the probability that n cars merge during a gap of t, and $E(t)$ the expected number of cars to merge during that gap, then by definition

$$E(t) = \sum_{n=0}^{\infty} n\alpha_n(t). \tag{85}$$

Thus, by our earlier remark, λ_c is given by

$$\lambda_c = \frac{1}{\mu} \int_0^{\infty} \varphi(t) E(t) \, dt, \tag{86}$$

where, as before, $\varphi(t)$ is the p.d.f. of the headways and μ is the mean headway.

The values of the $\alpha_n(t)$ can be calculated recursively from the relations

$$\alpha_0(t) = 1 - \alpha(t),$$

$$\alpha_n(t) = \alpha(t) \int_0^t g(\Delta)\alpha_{n-1}(t - \Delta)\, d\Delta \qquad n \geq 1. \tag{87}$$

The second line is derived by allowing the first car in line to merge with a move-up time Δ, followed by a merge of exactly $n - 1$ cars. A single integral equation for $E(t)$ can be derived from Eq. 87 as

$$E(t) = \alpha(t) \int_0^t g(\Delta)\, d\Delta + \alpha(t) \int_0^t g(\Delta)E(t - \Delta)\, d\Delta. \tag{88}$$

This equation does not have a simple solution in general, but in the special case of a constant move-up time Δ_0 we have $g(\Delta) = \delta(\Delta - \Delta_0)$, where $\delta(x)$ is a Dirac delta function, and Eq. 88 becomes

$$E(t) = 0 \qquad\qquad\qquad\quad t < \Delta_0,$$

$$E(t) = \alpha(t) + \alpha(t)E(t - \Delta_0) \qquad t > \Delta_0. \tag{89}$$

This equation has the solution

$$E(t) = \sum_{n=0}^{\infty} \alpha(t)\alpha(t - \Delta_0)\alpha(t - 2\Delta_0) \cdots \alpha(t - n\Delta_0), \tag{90}$$

where $\alpha(x)$ is to be set equal to zero if its argument is negative. In particular, if $\alpha(t) = H(t - T)$, then

$$E(t) = n + 1 \qquad \text{for} \qquad n\Delta_0 + T \leq t \leq (n + 1)\, \Delta_0 + T$$

$$= 0 \qquad \text{otherwise.} \tag{91}$$

This result implies that λ_c is

$$\lambda_c = \frac{1}{\mu} \sum_{n=0}^{\infty} (n + 1) \int_{n\Delta_0+T}^{(n+1)\Delta_0+T} \varphi(x)\, dx$$

$$= \frac{1}{\mu} \sum_{n=0}^{\infty} \phi(n\Delta_0 + T), \tag{92}$$

where $\phi(t) = \int_t^{\infty} \varphi(x)\, dx$. In particular, when $\phi(t) = \exp(-t/\mu)$ we find

$$\lambda_c = \frac{1}{\mu} \frac{\exp(-T/\mu)}{1 - \exp(-\Delta_0/\mu)}. \tag{93}$$

The parameter λ_c can be found experimentally if one knows the input rate on the feeder road λ and the probability that there are no cars in queue p_0 (on the assumption that the queue remains finite as $t \to \infty$). The following technique was suggested by Oliver.[12] The asymptotic expected number of cars to merge in time t is $\lambda_c(1 - p_0)t + o(t)$ for large t. The expected number of cars to arrive at the merge point during this same time is λt. Hence, by equating results in the limit $t = \infty$ we find the result

$$\frac{\lambda_c}{\lambda} = \frac{1}{1 - p_0}, \tag{94}$$

so that λ_c can be estimated from information on situations with relatively low flow.

The results above are just about the only generalities known for the merging–queueing problem; all other information has been derived for specific models. Furthermore, there is no one single model to which one can attribute greatest generality since all of the models treat some features of the queueing situation in detail and either ignore others or approximate them. Since considerable analysis is required to obtain detailed results for any of the models, we will not present the theoretical development here, but rather try to discuss the assumptions and differences inherent in each of the queueing models. The input process in the minor stream is generally assumed to be Poisson with intensity parameter denoted by λ. Furthermore, most authors use the assumption of a step gap acceptance function.

The first vehicular queueing model appears to have been one analyzed by Tanner.[54] This model deals with the queues generated by two conflicting traffic streams that compete for the use of a stretch of road AB wide enough for the use of a single vehicle only. Such a situation might arise at a narrow bridge or at a bottleneck in a two-lane highway. It is assumed that traffic stream 1 arrives at point A as a Poisson process with rate parameter λ_1 and that traffic stream 2 arrives at point B with rate parameter λ_2. The traversal time for a vehicle of type j $(j = 1, 2)$ to cross AB is a constant T_j. It is further assumed that there is a minimum time gap Δ_j between the passage of two successive vehicles of the same type through AB. Tanner distinguishes three cases:

(1) $T_1 > \Delta_1, \qquad T_2 > \Delta_2,$

(2) $T_1 > \Delta_1, \qquad T_2 < \Delta_2,$

(3) $T_1 < \Delta_1, \qquad T_2 < \Delta_2.$

Of these only the first is treated and, indeed, is the most interesting situation from the point of view of applications. Case 1 corresponds to a situation in which either one queue or the other has control of AB and retains control so long as vehicles remain in that queue. Cases 2 and 3 allow intermittent changes without the intervention of any external controls. The method of solution is a rather complicated version of the technique of embedded Markov chains, and the results are quite complicated even for the special cases for which they are derived. For example, when $\Delta_1 = \Delta_2 = 0$, that is, when there is no minimum gap between successive vehicles on AB, the mean delay to a vehicle in traffic stream 1 is

$$\bar{t}_1 = \frac{\exp{(\lambda_2 T_1)}\{\lambda_1 + \lambda_2 \exp{[(\lambda_1 + \lambda_2)T_1]}\}}{\lambda_2 \exp{(\lambda_2 T_1)}\{\exp{[\lambda_1(T_1 + T_2)]} - \exp{(\lambda_1 T_2)} + 1\} +}$$

$$+ \lambda_1 \exp{(\lambda_1 T_2)}\{\exp{[\lambda_2(T_1 + T_2)]} - \exp{(\lambda_2 T_1)} + 1\}$$

$$\times \frac{\exp{(\lambda_2 T_2)} - \lambda_2 T_2 - 1}{\lambda_2}, \quad (95)$$

with a similar result for $\bar{t}_2$ when the subscripts are interchanged. Tanner's model has recently been used by Blumenfeld and Weiss[39] to study the effects of splitting a queue to a major traffic stream to reduce the queueing time.

Hawkes[55,56] has discussed a slightly different version of this problem in which the time required to cross the stretch of road AB is a random variable with different distributions for the two streams. The analysis proceeds through the derivation of a coupled pair of integro–differential equations for the joint distributions of virtual waiting times[57] and leads to an expression for the mean delay when the system is in equilibrium. If μ_i and ν_i are the first and second moments of delay to an isolated car crossing the intersection, and $\rho_i = \lambda_i \mu_i$, then Hawkes' generalization of Eq. 95 is

$$\bar{t}_1 = \frac{\lambda_1 \nu_1[(1 - \rho_1) - 2\rho_2(1 - \rho_1 - \rho_2)] + \lambda_2 \nu_2(1 - \rho_1)}{2(1 - \rho_1 - \rho_2)(1 - \rho_1 - \rho_2 + 2\rho_1\rho_2)}. \quad (96)$$

Neither of the models just described is an exact generalization of the intersection delay problem that has been developed in earlier sections. In the notation of Tanner,[54] what is required is an analysis of the case $T_1 > \Delta_1$; $T_2 = 0$, $\Delta_2 > 0$, which is not amenable to the mathematical techniques of that paper. A later paper[58] contains an analysis of the merging–queueing problem with $\alpha(G)$ a step function. The precise assumptions are that cars on the main road can pass through the intersection in time T_1, and cars on the feeder road pass through in time T_2. Further, a car on the feeder road cannot enter the junction within a time Δ of a car in the main stream. Thus, Δ is the critical gap in a step gap acceptance function, and T_2 is the move-up time. The assumption of Poisson traffic is made for the major and minor roads with rate parameters λ_1 and λ_2, respectively. The result of greatest interest for which a reasonable simple expression can be given is that for expected delay when $T_1 = 0$. This corresponds to no queue in the major stream. The expected delay to a driver in the minor stream is

$$\bar{t} = \frac{\begin{aligned}&\lambda_1 \exp(\lambda_1 T_2)[\exp(\lambda_1 \Delta) - \lambda_1 \Delta - 1]\\ &\quad + \lambda_2 \exp(\lambda_1 \Delta)[\exp(\lambda_1 T_2) - \lambda_1 T_2 - 1]\end{aligned}}{\lambda_1\{\lambda_1 \exp(\lambda_1 T_2) - \lambda_2 \exp(\lambda_1 \Delta)[\exp(\lambda_1 T_2) - 1]\}}. \quad (97)$$

When $\lambda_2 = 0$ this reduces to the expected delay to a single driver at a merging point. The average queue length $\bar{n}$ for this model is just

$$\bar{n} = \lambda_2 \bar{t}. \quad (98)$$

A somewhat more complicated expression for $\bar{t}$ is available for $T_1 \neq 0$. Gaver[59] considered a queueing model similar to Tanner's, but under the restriction $T_2 = \Delta_2$, that is, the service time for a single vehicle is exactly equal to the critical gap. However, this requirement does not appear to be consistent with actual driver behavior. According to Hawkes,[34] Gaver's model can be modified to correct for this deficiency.

One of the simplest queueing models from the point of view of the analysis involved is that due to Buckley and Blunden,[60] who made use of known results for the $M/G/1$ queue. Although the authors claim considerable generality for their analysis, it appears to be correct only for Poisson traffic on the main road and a step gap acceptance function. The analysis makes use of results obtained earlier by Yeo[61] for an $M/G/1$ queue, in which an arriving customer at a queue has a different service time distribution depending on whether or not there are customers already in the queue. Most of the analysis required for this model involves the calculation of the statistics of delay for a driver at the head of queue (i.e., the service time), which is the same type of calculation given earlier in Section IV. The error in their analysis, except under conditions noted above, consists in assuming that successive headways are independent random variables. A residual gap is, however, not independent of total gap size when it is known that a vehicle has already accepted the gap.

Yeo and Weesakul[33] and Evans, Herman, and Weiss[32] have treated variants of Tanner's models, in which the merging time is related to the critical gap appearing in a step gap acceptance function. In these analyses a value of merging time is chosen from an arbitrary distribution, and this time is the critical gap in a step gap acceptance function. Thus, both of these models take into account the possibility of a distribution of critical gaps over the population. The expression obtained by Evans, Weiss, and Herman for the mean waiting time $\bar{t}$ is a relatively simple one. Let $g(t)$ be the probability density function for the merging time, and let $g^*(s)$ be its Laplace transform. Let the input of cars on the feeder road be Poisson with rate parameter λ, and let the probability density for headways be $\varphi(G) = \rho \exp(-\rho G)$.

Finally, define a dimensionless rate parameter β by $\beta = \lambda/(\lambda + \rho)$. The mean delay $\bar{t}$ is given by

$$\bar{t} = \left(\frac{1}{\rho}\right)\frac{1}{g^*(\rho)[g^*(\rho) - \beta]}\left(g^*(\rho) - [g^*(\rho)]^2 + \beta\rho\frac{dg^*}{d\rho}\right), \qquad (99)$$

where $g^*(\rho) > \beta$ for stationarity.

In particular, if $g(t) = \delta(t - T)$, that is, the merging time is a constant for the population, then

$$\bar{t} = \frac{1}{\rho}\left(\frac{1 - e^{-\rho T} - \beta\rho T}{e^{-\rho T} - \beta}\right). \qquad (100)$$

The mean queue length can be found through the use of Eq. 98. Possibly the most interesting conclusion in Yeo and Weesakul's study is that for low traffic densities the replacement of a distribution of a general form for $g(\tau)$ by a $g(t) = \delta(t - T)$ is a good approximation, while such an approximation leads to poor agreement with exact results at high densities. Hawkes[55] has also analyzed a model similar to those of Tanner, in which the major traffic stream is described as a succession of blocks and gaps, the blocks having a general distribution and the gaps having a negative exponential distribution. The gap acceptance function for the model is a step function.

The only queueing model that appears to allow correctly for a general gap acceptance function is one due to Hawkes.[34] In this analysis the traffic on the main road is described in terms of blocks and gaps, the blocks having a general distribution while the gaps have a negative exponential distribution. It is assumed that vehicles can merge with the main traffic stream only during a gap, but the merge takes place only with probability $\alpha(G)$ for a gap of duration* G. The function $\alpha(t)$ is assumed to satisfy the conditions

$$\begin{aligned}\alpha(G) &= 0 \qquad G < T_1, \\ \alpha(G) &= 1 \qquad G > T_1 + \delta,\end{aligned} \qquad (101)$$

where T_1 and δ are constants. Although this assumption disagrees with customarily assumed forms for $\alpha(G)$ except for the step function, it is undoubtedly correct. It is further assumed that the move-up time is a random variable M which satisfies $\delta \leq M \leq T$, where δ and T_1 are defined above. This inequality is somewhat questionable in the light of experimental values of $\alpha(G)$. Published experiments (e.g., Drew and co-workers[22,23] indicate values of T_1 of approximately 2 sec and δ of 3–5 sec. This contradicts the inequality $T_1 \geq \delta$. However, these measurements represent mean values of $\alpha(G)$

---

* It should be pointed out that this represents a slight difference in terminology between this work of Hawkes and that of other workers.

averaged over a population of drivers rather than measurements on individuals, for which Hawkes' theory is more appropriate. The measurements of Herman and Weiss[19] on individuals lead to values in conformity with Hawkes' assumptions ($T_1 \sim 3$ sec, $\delta \sim 2$ sec), but more experimental values would be required to validate the assumption. Hawkes' paper contains further minor assumptions related to what he calls the hesitation time H. This time is relevant for vehicles arriving at an empty queue. It is the time before the driver begins to search for a gap. One can realistically set $H = 0$, so that the parameter is not likely to be critical in applications of the theory. The mathematical development uses the theory of Yeo[61] referred to above, and leads to an expression for the Laplace transform of the probability density for the average waiting time. Hawkes reported on some calculations of mean delay based on a step gap acceptance function with the critical gap uniformly distributed around the mean. For this situation he found that the theory with a fixed step gap acceptance function reproduces the theoretical results quite acceptably, lending support to similar conclusions by Evans, Weiss, and Herman[32] and Yeo and Weesakul.[33]

There is very little in the queueing literature involving multiple lane crossing. One analysis of some interest is that of Hawkes.[62] His model involves the delay to a queue of cars, some waiting to turn into one lane on the major road, the remainder waiting either to join the other lane or to cross both lanes. The probability of a given driver belonging to one category or the other is p_i ($i = 1, 2$), with $p_1 + p_2 = 1$. The headway distribution in either lane is negative exponential, with possibly different flow rates. The gap acceptance function is of the form

$$\alpha(G_1, G_2) = H(G_1 - T_1)H(G_2 - T_2), \tag{102}$$

where $H(x)$ is a step function, and the critical gaps in the near and far lanes satisfy $a \leq T_1 \leq T_2 \leq 2a$, where a is a constant. In practice this does not appear to be a serious restriction.* The analysis is fairly simple, but the results are quite involved and will not be reproduced here. As a special case Hawkes treats the problem of left turners† in a two-lane traffic, generalizing some work by Newell.[63]

Some recent work has appeared in the literature by Reid[64] on the delay to vehicles on a two-lane road, in which left-turning vehicles‡ can cause

---

* Some unpublished measurements of Herman and Weiss indicate that single driver gap acceptance functions for two-lane traffic can be approximated by Eq. 102 with values of T_1 and T_2 differing by not more than 1 sec. Since a would lie between 1.5 and 3 sec, this would confirm the validity of Hawkes' assumption.

† Right turners in Hawkes' terminology, since the paper is written for an English audience.

‡ Right-turning vehicles in the original terminology.

delays to left-turning vehicles in the oncoming stream of traffic at an un-controlled intersection. Vehicles are classified as being of type 1 or 2 according to whether they proceed straight through the intersection or plan to turn, and a given vehicle is of type i with probability p_i. Acceptance of a gap by a type 2 vehicle depends on the choice of a crossing time τ sampled from arbitrary distributions if the opposing vehicle (in the opposite lane) is type 1. A crossing is made immediately if the opposing vehicle is type 2. If the gap is greater than τ, the driver crosses in time τ, while if it is less, he waits and becomes the head of a queue. A type 2 vehicle just arriving at the head of a queue makes a decision to cross immediately, with probability R_i (i refers to the lane) if the opposing vehicle is of type 1. These last two assumptions are both somewhat unrealistic, particularly the second, which appears to take no account of the size of gap in the opposing traffic stream. Hence, a good theory of the effects of left turners in the two-stream situation does not appear to be available at present.

Finally, mention should be made of queueing models by Winsten[13] and by Oliver and Bisbee,[65] since these are sometimes cited in the literature. The first of these involves queueing in discretized time, so that the underlying Markov process can be replaced by a Markov chain, and the second of these is a high flow model in which at most one car in a queue on the side road can accept a given gap. Since both models have been superseded by later work, no more detailed description will be given.

The preceding discussion should suffice to indicate the present state of queueing problems for merging and crossing situations. There are many theoretical treatments of related problems. It is unclear at the present time (1973) whether the theories are significantly different, or lead to results that are essentially the same. There are no experimental results which bear directly on the queueing situation. In general, one can say that it would be difficult at the present time for a traffic engineer to make use of any but the simplest results available if he requires any queueing considerations in his work.

VII. THE DELAY AT A TRAFFIC SIGNAL

A great deal of the literature in the mathematical theory of traffic flow has been devoted to the analysis of motorist delays at an intersection controlled by a fixed-cycle traffic signal. The problem—to determine the signal settings that minimize the expected delay to the motorist—is of obvious practical importance. The first problem to consider is that of finding the delay to a single stream of traffic that forms an input to the intersection. Since the intersection may be regarded as being composed of a number of input and output streams converging at a point, the overall delay optimization problem may be attempted once the delays for each stream for any given signal discipline are known.

In order to obtain an expression for the expected delay for a single stream, it is necessary to specify, besides the signal discipline, the arrival process for the incoming vehicles and the process describing the manner in which vehicles pass through the intersection.

The Arrival Process

Many attempts to obtain the expected delay rest on the assumption that, for each input stream, the vehicles arrive according to a simple Poisson process. As stated earlier, in Section II, the simple Poisson process seems to be a reasonable model if the traffic flow is light. Under less ideal conditions—in heavy traffic, for example, where the interactions between vehicles cannot be neglected—one would expect departures from the simple Poisson model, and the simplest generalization is the compound Poisson process in which, if $N(t)$ is the number of arrivals in any interval of length t,

$$E[z^{N(t)}] = \exp\{-\Lambda t[1 - \phi(z)]\}. \tag{103}$$

It follows from Eq. 103 that if $\lambda = \Lambda\phi'(1)$, then

$$E[N(t)] = \lambda t, \tag{104}$$

$$\mathrm{var}\,[N(t)] = I\lambda t, \tag{105}$$

where I, the *index of dispersion* of the process is given by

$$I = 1 + \frac{\phi''(1)}{\lambda}. \tag{106}$$

As a consequence, the index of dispersion for a compound Poisson process is never less than unity, its value for a simple Poisson process. Although some experimental studies (see, for example, Miller[66]) have indicated that departures from the simple Poisson stream in heavy traffic are in this direction, often with an index of dispersion of the order 2, it would be of interest to obtain a more general and mathematically tractable process in which the index of dispersion can be less than unity. One way of doing this is that adopted by Darroch,[78] and may be described as follows.

In the process of binomial arrivals described in Section II, let time be regarded as consisting of discrete, contiguous intervals, each of duration h. Let the numbers of arrivals in the h intervals be independent, identically distributed random variables with probability generating function $\varphi(z)$. Clearly this process specializes to the binomial process if

$$\varphi(z) = 1 - p + pz. \tag{107}$$

However, if

$$\varphi(z) = \exp\{-\lambda h[1 - \phi(z)]\}, \tag{108}$$

and we assume, in addition, that the arrivals during a particular h interval are randomly distributed over that interval, the process reduces to the compound Poisson process described above. Since the index of dispersion for the binomial and compound Poisson processes are, respectively, less and greater than unity, we see that the above process generalizes the simple Poisson process in both directions. In general it may be shown that the index of dispersion for Darroch's process is given by

$$I = 1 - \lambda h + \frac{\varphi''(1)}{(\lambda h)}. \tag{109}$$

The Departure Process and Signal Discipline

Let the period of the fixed-cycle traffic signal be T, and denote by R the "red period," that is, the part of the cycle during which time the signal is against the arriving stream of traffic under consideration. The interval R includes the "amber period" and any time lost due to change-over of signals. Thus it is assumed that there is no restriction on the departure of vehicles during the remaining "green period" of duration $T - R$. The simplest method of describing the manner in which vehicles depart from the intersection is to specify that when free to do so the vehicles depart in turn, the intervals between successive departures being independent and identically distributed random variables. In the case of a single lane of traffic it may be reasonable to assume that these departure times are constant. This assumption has been made in most of the literature on the fixed-cycle traffic signal problem. While one would expect the departure time of the first vehicle to be greater than those of the succeeding vehicles, this circumstance may be allowed for by increasing the value of R. It is also assumed that vehicles that arrive when there is no queue are not delayed, that is, have zero departure times.

If all the vehicles do not proceed straight across the intersection, but some turn right and some turn left, the departure times are by no means constant. The vehicles that turn right and thus do not have to cross oncoming traffic (left, in Britain and Australia and other countries in which motorists drive on the left-hand side of the road) may be considered to have greater departure times than those proceeding straight ahead. If left turns are prohibited, one could assume that the departure times have a distribution containing two jumps, one for each of the two classes of motorists. In any case it may be reasonable to assume that the departure times are independent, identically distributed random variables.

It is not so easy to dispose of the complication of vehicles wishing to turn across the face of oncoming traffic. If such motorists have a special lane in the input stream, then their delays may be considered separately to the delays of those in the main stream, irrespective of whether there is a special green period for them or not. If, however, there is no special lane for such vehicles, even the assumption of independent and identically distributed random departure times is unreasonable, since one motorist may block all those behind for the duration of the cycle. A theoretical analysis of delays due to vehicles turning across the face of traffic was attempted by Newell,[68] and such turning motorists (as well as interfering pedestrians) were allowed for in a simple model considered by Darroch.[78]

Gordon and Miller[69] also discussed this problem, and Little[50] evaluated delays to motorists wishing to make various turning maneuvers in Poisson traffic. Both of these investigations, however, are concerned only with the delays to the turning vehicles, and the more important problem of determining the expected delay caused by such vehicles is not considered. Reid[64] and others have considered this problem for uncontrolled intersections, but there is yet to appear a useful theoretical analysis of the expected delay caused by turning vehicles at the fixed-cycle signalized intersection.

The Delay

In queueing theory (see Feller[70] for a résumé) a random variable of interest is the delay to an individual. The classical queueing situation is simpler than the process describing the build-up of vehicles at a traffic signal, and may be described, in the simplest case, as follows. Individuals arrive at a certain point in such a way that the instants of arrival form a renewal process. The individuals are given service in the same order as that in which they arrive, and the service periods are independent, identically distributed random variables. No individual may depart until service is completed. There are three random variables of interest: the queueing time of an individual (the time spent waiting in the queue not counting the service time), the number of individuals in the system at the instants of time just after the departures of each individual, and the busy period (the interval of time during which there is at least one individual in the system). In the case when the distributions of either the interarrival intervals or the service periods have an exponential tail, the distributions or transforms thereof of all three random variables of interest are obtainable, but in general the evaluation of these distributions presents a difficult problem, further discussion of which would be out of place here.

The traffic signal situation is complicated by the fact that part of the time (during the red periods) no departures at all are possible. Another difference from the simple queueing model is that the vehicles at the traffic signal do not

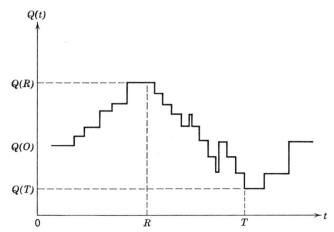

FIGURE 1. A realization of the queueing process at the traffic signal.

necessarily depart in order of arrival, although this is the case for a single lane of traffic. The random variable of most interest in the traffic signal situation is the total delay to all the vehicles during a cycle. There are two reasons for focusing attention on this random variable rather than on the individual delay: the individual wait seems difficult to obtain because of the above-mentioned differences between the traffic queue and the ordinary queueing situation, and the total delay seems to be of more importance in traffic flow theory, giving as it does a direct measure of the financial cost due to the congestion. In any case, some measure of the individual delay could be obtained by dividing the total delay for a cycle by the expected number of vehicles to arrive in a cycle. In particular, it would be expected that under very general conditions the mean total delay would be equal to the product of the expected individual delay and the expected number of arrivals.

Figure 1 depicts a realization of the process $Q(t)$, where $Q(t)$ is the number of vehicles in the traffic signal queue (including the one, if any, in the process of departing) at time t after the commencement of a red period. Clearly the total delay to those vehicles in the system during the interval $(t, t + \delta t)$ is $Q(t)\,\delta t + o(\delta t)$, so that the total delay during the cycle time T is

$$W = \int_0^T Q(t)\,dt.$$

The Mean Delay for Poisson Input and Constant Departure Times

Denote by $A(t)$ the number of vehicles which join the queue in the time interval $(0, t)$, the origin being taken at the start of the red period. No departures are possible in the interval $(0, R)$, the red period, but there is no restriction on departures in (R, T), the green period. We assume that $A(t)$ is a simple Poisson process with $E[A(t)] = \lambda t$, and that each motorist takes the time s to depart from the queue. Put

$$W = W_1 + W_2, \tag{110}$$

where

$$W_1 = \int_0^R [Q(0) + A(t)] \, dt, \tag{111}$$

$$W_2 = \int_R^T Q(t) \, dt. \tag{112}$$

Taking expectations in Eq. 111 we find

$$E[W_1] = RE[Q(0)] + \tfrac{1}{2}\lambda R^2. \tag{113}$$

To find $E[W_2]$, consider first the associated random variable W_2^*, defined as W_2 conditonal on $T = \infty$. W_2^* may be regarded as the total wait in a busy period for a queueing process $X(t)$ with simple Poisson arrivals, constant service times, and $X(0) = Q(R)$. It can be shown (see, for example, McNeil[71] and Daley and Jacobs[72]), that provided $\lambda s < 1$,

$$E[W_2^*] = \frac{sE[Q(R)]}{2(1 - \lambda s)^2} + \frac{sE[Q^2(R)]}{2(1 - \lambda s)}. \tag{114}$$

To avoid trivial algebraic complications, it is now necessary to assume that $T - R$ is an integer multiple of s, that is, the green period contains an integral number of departure times. Since W_2 is obtained from W_2^* by neglecting that part of the queueing process after time T, we have

$$E[W_2] = E[W_2^* \mid X(0) = Q(R)] - E[W_2^* \mid X(0) = Q(T)]$$

$$= \frac{sE[Q(R) - Q(T)]}{2(1 - \lambda s)^2} + \frac{sE[Q^2(R) - Q^2(T)]}{2(1 - \lambda s)}, \tag{115}$$

using Eq. 114.

Now let us assume that the queue is in statistical equilibrium. A necessary and sufficient condition for this is that the average number of arrivals per cycle be less than the number that can pass through in a green period, that is, $\lambda T < (T - R)s^{-1}$, or

$$\lambda s < 1 - \frac{R}{T}. \tag{116}$$

In this case $E[Q(0)] = E[Q(T)]$ and $E[Q^2(0)] = E[Q^2(T)]$. Also $Q(R) = Q(0) + A(R)$, so

$$E[Q(R) - Q(T)] = E[A(R)] = \lambda R, \tag{117}$$

$$E[Q^2(R) - Q^2(T)] = 2E[A(R)]E[Q(0)] + E[A^2(R)]$$

$$= 2\lambda RE[Q(0)] + \lambda^2 R^2 + \lambda R. \tag{118}$$

Equations 115, 117, and 118 yield

$$E[W_2] = \tfrac{1}{2}s(1 - \lambda s)^{-2}\{\lambda R + (1 - \lambda s)(2\lambda RE[Q(0)] + \lambda^2 R^2 + \lambda R)\}, \tag{119}$$

whence, using Eqs. 110 and 111, we get

$$E[W] = \frac{\lambda R}{2(1 - \lambda s)}\left[R + \frac{2}{\lambda}E[Q(0)] + s\left(1 + \frac{1}{1 - \lambda s}\right)\right] \tag{120}$$

for the expected total wait per cycle. If the right-hand side of Eq. 120 is divided by λT, the average number of vehicles to arrive in a cycle, we obtain the mean delay to an individual motorist.

If the traffic is sufficiently light, the term involving $E[Q(0)]$ may be neglected since it is unlikely that any vehicles will remain in the queue at the end of the green period. Otherwise an expression for $E[Q(0)]$ must be obtained before Eq. 120 is of practical value, and this is the problem of greatest difficulty in analyzing the delay at the fixed-cycle signalized intersection. Before discussing this problem, we will discuss methods of generalizing Eq. 120 to allow for more general arrival and departure processes.

The Mean Delay for a More General Arrival Process

The arguments leading to Eq. 120 can be generalized. Suppose that the arrival process is that postulated by Darroch, while the departure times remain constant. If we assume that arrivals occur at random during the h intervals, we have $E[A(t)] = \lambda t$, so Eq. 113 remains valid.

To obtain $E[W_2]$, algebraic complications arise unless s is a multiple of h, so we assume this to be the case. It can now be shown that Eq. 114 generalizes to

$$E[W_2] = \frac{s(1 - \lambda s + \lambda sI)E[Q(R)]}{2(1 - \lambda s)^2} + \frac{sE[Q^2(R)]}{2(1 - \lambda s)}, \tag{121}$$

where I is the index of dispersion of the arrival process $A(t)$. Consequently, Eq. 120 becomes

$$E[W] = \frac{\lambda R}{2(1 - \lambda s)}\left[R + \frac{2}{\lambda}E[Q(0)] + s\left(1 + \frac{I}{1 - \lambda s}\right)\right]. \tag{122}$$

Equation 122 is essentially that obtained by Darroch.[78] For binomial arrivals with $h = s$, Eq. 109 gives $I = 1 - \lambda s$, so Eq. 122 specializes to

$$E[W] = \frac{\lambda R}{2(1 - \lambda s)}\left(R + \frac{2}{\lambda} E[Q(0)] + 2s\right), \tag{123}$$

which is the formula obtained by Beckmann et al.[13] (In fact, the formula of Beckmann et al. contains an s instead of a $2s$ in the right-hand side of Eq. 123, but in their model it is assumed that the first departure occurs at time $\frac{1}{2}s$ after the start of a given period, whereas in the model used to obtain Eq. 122 the first departure occurs after a time s. The two models give the same result if the value of $E[Q(0)]$ in their case is increased by $\frac{1}{2}\lambda s$, which is just the average number of arrivals in half an h interval. This is exactly what we would expect.)

Consider now the simplest arrival process of all, namely, the situation where the vehicles in the input stream are each separated by the constant interval λ^{-1}. This simple model was first investigated by Clayton,[73] who used it to obtain the signal settings which minimize the delay at an intersection. While the process $Q(t)$ is now completely deterministic and thus trivial, it is worth considering since it gives a measure of the minimum possible delay in the sense that the amount by which the expected delay is greater than this minimum value may be regarded as due to the randomness in the input process. Clearly in this case $Q(0) \equiv 0$ and $I = 0$, so Eq. 122 becomes

$$E[W] = \frac{\lambda R}{2(1 - \lambda s)}(R + s), \tag{124}$$

the formula obtained by Clayton. Comparing this to Eq. 122, we may regard the second term in the right-hand side of Eq. 122 as arising from the possibility of overflow from one cycle to the next, while the fourth term arises from the randomness in the arrival process within a single cycle. The third term, s, could be regarded as arising from the initial inertia in the departure process and is not due to any randomness.

Except for Newell's work,[74] which deals with approximation methods for the near-saturation situation, no one has considered the problem of determining $E[W]$ for inputs other than those which are a special case of Darroch's process. This is not surprising, for even in the simple queueing situation analytic results are not known for general independent interarrival periods and constant service times. It may be possible to obtain results for a general renewal arrival process if we assume that the departure times are not constant, but negative exponential random variables, since this would correspond to a simple queueing situation for which the individual delay distribution is explicitly obtainable.

The Mean Delay for Random Departure Times

In some situations, if some drivers wish to make right turns (as distinct from turns across the face of oncoming traffic), for example, the assumption that departures occur after constant intervals of time is unreasonable. More generally, let us suppose that the departure times are independent, identically distributed random variables, having mean value s and coefficient of variation C. For the arrival process we specialize to the simple Poisson process.

For this model the expected total delay during the red period is, of course, the same as for the model with constant departure times, and is therefore given by Eq. 113. To find $E[W_2]$ we can use a result of Daley and Jacobs,[72] which generalizes Eq. 113 to

$$E[W_2^*] = \frac{s(1 + \lambda s C^2)E[Q(R)]}{2(1 - \lambda s)^2} + \frac{sE[Q^2(R)]}{2(1 - \lambda s)}. \tag{125}$$

However, the expression for $E[W_2]$ is not obtained simply by replacing $Q(R)$ by $Q(T)$ in Eq. 125 and subtracting, since the green period commences with a departure period but does not, in general, end with one (unless the last departure time in the green period is cut short to allow the last vehicle to escape, in which case the assumption of identically distributed departure times is violated). It may be reasonable to assume that the end of the green period is positioned independently of the process of departures, in which case $E[W_2]$ may be determined as follows: If we replace $Q(R)$ by $Q(T)$ in Eq. 125, we obtain a value which is greater than what we want by the amount $\Delta E[Q(T)]$, where Δ is the expected difference between a departure time that starts at time T and the amount by which a departure time that covers the time epoch T overshoots the end of the green period. Thus

$$\Delta = s - E[V],$$

where V is the forward recurrence time in the departure process. Using a result in renewal theory (see, for example, Cox,[2] p. 63) we have

$$E[V] = \tfrac{1}{2}s(1 + C^2),$$

whence

$$\Delta = \tfrac{1}{2}s(1 - C^2). \tag{126}$$

Using Eq. 125 we find

$$E[W_2] = \frac{s(1 + \lambda s C^2)E[Q(R) - Q(T)]}{2(1 - \lambda s)^2} + \frac{sE[Q^2(R) - Q^2(T)]}{2(1 - \lambda s)} + \Delta E[Q(T)].$$

This, together with Eqs. 113, 117, 118, and 126 yields

$$E[W] = \frac{\lambda R}{2(1 - \lambda s)}\left[R + \frac{2}{\lambda}\left(1 + \frac{(1 - \lambda s)(1 - C^2)s}{2}E[Q(0)]\right.\right.$$
$$\left.\left. + s\left(1 + \frac{1 + \lambda s C^2}{1 - \lambda s}\right)\right)\right]. \tag{127}$$

If the arrival process is compound Poisson, it is easily shown that the only change to Eq. 127 is the replacement of the term $1 + \lambda s C^2$ in the right-hand side by $I + \lambda s C^2$. We are now able to drop the trivial requirement that, for the traffic signal with constant departure times s, the green period be an integral multiple of s. This result is given by Eq. 127 with $C = 0$. Except for the slight increase (or decrease, if $C > 1$) in the coefficient of the overflow $E[Q(0)]$, Eq. 127 tells us that the effect of the randomness in the departure process is the same as would be obtained by increasing the index of dispersion by $\lambda s C^2$, where C is the coefficient of variation for the departure times.

In some situations it may be reasonable to assume that the departure times are negative exponentially distributed. This would be approximately the case if the output stream consisted of several lanes, with the departure processes for the various lanes independent. Besides simplifying Eq. 127 (the correction term to $E[Q(0)]$ vanishes since $C = 1$, this being a consequence of the no-memory property of the exponential distribution), the process of negative exponential departures has the added attraction of enabling us to generalize the arrival process. The most elegant results in queueing theory relate to the simple queue with negative exponential service times and general, identically distributed interarrival periods. As we would expect, it is also possible to analyze the traffic signal delay for this model of arrivals and departures by writing down and solving a system of difference equations for $E[W_i]$, where W_i is the total delay during a busy period commencing with the simultaneous arrival of i individuals (see Daley[75]).

The Overflow at the Traffic Signal

Haight[76] studied the distribution of $Q(0)$ for the traffic signal with simple Poisson input and constant departure times. These results are of limited usefulness in the present context, however, since what Haight obtained is the distribution of $Q(0)$ conditional on the number in the queue at the end of the previous red period, and the latter random variable is just as difficult to analyze as $Q(0)$. For the case of constant departure times, exact expressions for $E[Q(0)]$ are obtainable in terms of the complex roots of certain transcendental equations. For a binomial input process put $h = s$ and suppose that the red period consists of $q = R/s$ intervals, each of length s (the departure time for a vehicle) and that the green period consists of $m = (T - R)/s$ such intervals. Let Y_k denote the number of arrivals in the interval $(ks, ks + s)$. Then considering transitions from the end of an arbitrary green period to the end of the next green period, we have

$$Q(T) = \max \left(Q(0) + \sum_{k=0}^{q+m-1} Y_k - m, 0 \right). \tag{128}$$

In the language of queueing theory, Eq. 128 represents a queue with constant service times of length T, service in batches of m (or less, if there are fewer than m individuals waiting), and arrivals such that the numbers of arrivals in each service period are independent random variables with probability generating function $(1 - \lambda s + \lambda s z)^{q+m}$. There is the additional requirement that service shall proceed even if the system is empty, and $Q(T)$ represents the number waiting just after the commencement of an arbitrary service period.

The above bulk service queue was first studied in the traffic theory context by Newell,[77] who used generating function techniques to obtain $\Psi(z)$, where

$$\Psi(z) = E\{z^{Q(0)}\} = \sum_{n=0}^{\infty} z^n \, \text{Prob} \, \{Q(0) = n\}. \tag{129}$$

Newell's result is

$$\Psi(z) = \prod_{k=1}^{q} \left(\frac{1 - z_k}{z - z_k} \right), \tag{130}$$

where $z_1, z_2, \ldots, z_q$ are the roots of the equation

$$z^m - (1 - \lambda s + \lambda s z)^{q+m} = 0 \tag{131}$$

outside the unit circle $|z| = 1$. It follows from Eq. 130 that the overflow is given by

$$E[Q(0)] = \Psi'(1) = \sum_{k=1}^{q} \frac{1}{z_k - 1}. \tag{132}$$

Newell was unable to obtain explicit expressions for these roots, except in the special cases $q/m = 1, \frac{1}{2}, 2, \frac{1}{3}$, and 3, when Eq. 131 reduces to a set of quadratic, cubic, or quartic equations. In the case $q/m = 1$, for example (which corresponds to a cycle time that is twice the effective green time), Eq. 131 becomes

$$(\lambda s)^2 z^2 + \left[2\lambda s(1 - \lambda s) - \exp\left(\frac{2\pi i k}{q}\right) \right] z + (1 - \lambda s)^2 = 0, \tag{133}$$

$k = 0, 1, 2, \ldots, q - 1$, and each of these quadratic equations has a single root outside the unit circle, namely,

$$z_k = -\left(\frac{1 - \lambda s}{\lambda s} \right) + \frac{1}{2}(\lambda s)^{-2}$$

$$\times \left\{ 1 + \left[1 - 4\lambda s(1 - \lambda s) \exp\left(\frac{-2\pi i k}{q}\right) \right]^{1/2} \right\} \exp\left(\frac{2\pi i k}{q}\right). \tag{134}$$

For inputs other than the simple binomial process, the overflow at the traffic signal is no longer given by the queueing (Eq. 128). In general, it is possible for *more* than m vehicles to be discharged during a green period. This would happen if $Q(t)$ decreases to zero at time ks, say, after the commencement of the green period, and $m - k + 1$ or more vehicles arrive before the green period ends. To allow for this possibility, it will be necessary to write down an equation relating the values of $Q(t)$ at the successive departure points, not just at the beginning and end of the cycle. Following Darroch[78] we have, for $k = q, q + 1, \ldots, q + m - 1$ (i.e., during the green period),

$$Q((k + 1)s) = Q(ks) + Y_k - 1 \quad \text{if} \quad Q(ks) > 0,$$
$$0 \quad \text{if} \quad Q(ks) = 0, \tag{135}$$

where Y_k is the number of arrivals in the kth s interval of the cycle. Denote by $\varphi(z)$ the probability generating function of Y_k and by $\Psi_k(z)$ the probability generating function of $Q(ks)$. Using classical probability generating function techniques we obtain, directly from Eq. 135,

$$\Psi_{k+1}(z) = \left[\frac{\varphi(z)}{z}\right][\Psi_k(z) - \Psi_k(0)] + \Psi_k(0),$$

and therefore

$$\Psi_{q+m}(z) = \left(\frac{\varphi(z)}{z}\right)^{q+m} \Psi_q(z) + \left(\frac{1 - \varphi(z)}{z}\right)^{q+m-1} \sum_{k=q}^{} \Psi_k(0)\left(\frac{z}{\varphi(z)}\right)^{k-q-m+1}. \tag{136}$$

Considering the red period, we have $\Psi_q(z) = \{\varphi(z)\}^q \Psi(z)$. Also $\Psi(z) = \Psi_{q+m}(z)$ in equilibrium. Putting these relations in Eq. 136 we get, for the probability generating function of $Q(0)$,

$$\Psi(z) = \left(\frac{(\zeta - 1)\{\varphi(z)\}^m}{z^m - \{\varphi(z)\}^{q+m}}\right) \sum_{k=0}^{m-1} p_{k+q} \zeta^k, \tag{137}$$

where $\zeta = z/\varphi(z)$ and p_k is the probability that $Q(ks)$ is zero. Using the fact that $\Psi(z)$ is a probability generating function and, therefore must converge within the unit circle $|z| = 1$, Darroch showed that Eq. 137 can be written

$$\Psi(z) = \left(\frac{m - (q + m)\lambda s}{1 - \lambda s}\right)\left(\frac{(\zeta - 1)\{\varphi(z)\}^m}{z^m - \{\varphi(z)\}^{q+m}}\right) \prod_{k=1}^{m-1} \left(\frac{\zeta - \zeta_k}{1 - \zeta_k}\right), \tag{138}$$

where $\zeta_k = \zeta(z_k) = z_k/\varphi(z_k)$ and $z_1, z_2, \ldots, z_{m-1}$ are the $(m - 1)$ zeros of $z^m - \{\varphi(z)\}^{q+m}$ that are strictly inside the unit circle. The expectation of $Q(0)$ is now obtained by differentiating Eq. 138. We deduce the relation

$$E[Q(0)] = (1 - \lambda s) \sum_{k=1}^{m-1} \frac{1}{1 - \zeta_k}$$
$$+ \left(\frac{m - (q + m)\lambda s}{1 - \lambda s}\right)\frac{d}{dz}\left(\frac{z\varphi^{m-1}(z) - \varphi^m(z)}{z^m - \varphi^{q+m}(z)}\right)_{z=1}. \tag{139}$$

The problem remains to obtain an expression for $\sum (1 - \zeta_k)^{-1}$. Darroch did not attempt to do this; instead, he obtained bounds for $E[Q(0)]$ by proceeding directly from Eq. 137. It is in fact possible to obtain an explicit formula for $E[Q(0)]$ in the case when $\varphi(z)$ does not have any zeros within the unit circle. The method, involving contour integration, was originally employed by Crommelin[79] to obtain the expected waiting time in the many-server queue with simple Poisson arrivals and constant service times, and was also used by McNeil[71] to obtain a close upper bound for $E[Q(0)]$ for the traffic signal with compound Poisson arrivals. An exact, though complicated, series expansion for $E[Q(0)]$ for simple Poisson arrivals was obtained by Kleinecke.[80]

The difficulty of obtaining simple, easily computable expressions for $E[Q(0)]$ has prompted many authors to look for approximations and bounds. A valid, though trivial, lower bound for $E[Q(0)]$ is zero. This is useful if the traffic intensity is small; in fact, Miller[81] suggested that $E[Q(0)]$ is not much different from zero as long as the normalized traffic intensity $\lambda s/[1 - (R/T)]$ is smaller than 0.5. In the same paper Miller derived an upper bound for $E[Q(0)]$ that improves as the traffic intensity increases. For constant departure times the queueing equation (Eq. 128) is accurate for the traffic signal for binomial arrivals with $h = s$, but in general the value it gives for $E[Q(0)]$ clearly always exceeds the true value. For general independent and identically distributed interarrival and departure times, the queueing Eq. 128 generalizes to

$$Q^*(T) = Q^*(0) + A(T) - D(G) + \delta, \tag{140}$$

where $A(T)$ is the number of arrivals and $D(G)$ the number of departures (if they were available) in the cycle, and $D(G) - \delta$ is the number which actually depart in this period. For the traffic signal, $A(T)$ may include arrivals which are not delayed, so $Q^*(T)$ is stochastically larger than the random variable $Q(T)$ corresponding to the traffic signal problem.

Following Miller,[81] we take expectations in Eq. 140 to obtain, in equilibrium,

$$E[\delta] = E[D(G) - A(T)]. \tag{141}$$

Next we rewrite Eq. 140 as

$$Q^*(T) - \{\delta - E[\delta]\} = Q^*(0) - \{D(G) - A(T) - E[D(G) - A(T)]\}.$$

Squaring both sides, and again taking expectations, we find

$$E[Q^*(T)]^2 + 2E[Q^*(T)]E[\delta] + \text{var } [\delta]$$
$$= E[Q^*(0)]^2 + \text{var } [\delta - A(T)]. \tag{142}$$

Here we have used the fact that the product $Q^*(T) \delta$ is always zero. Using Eq. 141 and the equilibrium relation $E[Q^*(T)]^2 = E[Q^*(0)]^2$, we deduce from Eq. 142 that

$$E[Q^*(0)] = \frac{\text{var}\,[D(G) - A(T)] - \text{var}\,[\delta]}{2E[D(G) - A(T)]}. \tag{143}$$

The second term in the numerator of Eq. 143 is now the only unknown, but as the traffic intensity increases, var $[\delta] \to 0$. In any case

$$E[Q(0)] \le E[Q^*(0)] \le \frac{\text{var}\,[D(G) - A(T)]}{2E[D(G) - A(T)]}.$$

For Darroch's arrival process and general independent and identically distributed departure times, $D(G) = (T - R)/s$, $E[A(T)] = \lambda T$ and var $[A(T)] = I\lambda T$, so this becomes

$$E[Q(0)] \le \frac{I\lambda s}{2(1 - R/T - \lambda s)}. \tag{144}$$

An exact expression for $E[Q^*(0)]$ was obtained by McNeil.[71] In the case of simple Poisson arrivals and constant departure times it is

$$E[Q^*(0)] = \sum_{n=1}^{\infty} \frac{1}{n} \sum_{j=1}^{\infty} \frac{j}{(j + mn)!}\,(n\lambda T)^{j+mn}, \tag{145}$$

where $m = (T - R)/s$. This series converges rapidly unless λs is near $1 - R/T$ (λs must be less than $1 - R/T$ for equilibrium), and provides a much closer upper bound to the overflow to be reached than the right-hand side of Eq. 144. A lower bound for $E[W]$ is obtained by putting $E[Q(0)]$ equal to zero in Eq. 127, and upper bounds are obtained by replacing $E[Q(0)]$ by $E[Q^*(0)]$ and the right-hand side of Eq. 144, respectively.

In fact (and this seems to have been overlooked by many people), it is possible to obtain $E[Q(0)]$ exactly, using standard numerical techniques. Consider, for example, the case of simple Poisson arrivals and constant departure times. In this case, $\varphi(z) = \exp\{-\lambda s(1 - z)\}$ and Eq. 139 becomes

$$E[Q(0)] = (1 - \lambda s) \sum_{k=1}^{m-1} \frac{1}{1 - \zeta_k} + \tfrac{1}{2}(1 - \lambda s)$$

$$+ \frac{\lambda s(R/T)}{2(1 - \lambda s)(1 - R/T - \lambda s)} - \frac{1}{2}\left\{\frac{T}{s}\left[1 - \rho - \frac{R}{T} - 2\rho\frac{R}{T}\right]\right\}, \tag{146}$$

where each ζ_k is the unique root within the unit circle of

$$\zeta_k \exp\left(\frac{\lambda s(1 - \zeta_k)}{(1 - R/T)}\right) = \exp\left(\frac{2\pi i k}{m}\right). \tag{147}$$

TABLE 3

VALUES OF BOUNDS FOR $E[W]/(\lambda T^2)$

T/S		λs		
		0.165	0.33	0.495
20	L	0.1826	0.2331	0.3213
	E	0.1828	0.2492	5.093
	U_1	0.1828	0.2505	5.104
	U_2	0.2274	0.3428	5.271
40	L	0.1661	0.2099	0.2845
	E	0.1662	0.2136	2.636
	U_1	0.1662	0.2138	2.640
	U_2	0.1885	0.2646	2.760
80	L	0.1579	0.1983	0.2660
	E	0.1579	0.1986	1.416
	U_1	0.1579	0.1986	1.418
	U_2	0.1691	0.2256	1.504
∞	L	0.1497	0.1866	0.2475
	U_2	0.1497	0.1866	0.2475

Siskind[82] has pointed out that many computer routines for solving such equations are available.

Values of the bounds for $E[W]/(\lambda T^2)$, the mean delay per vehicle per unit cycle time, are given in Table 3 for the case $R = \frac{1}{2}T$. L is the lower bound obtained by neglecting the overflow, E the exact value using Eqs. 146 and 147, and U_1 and U_2 are the upper bounds given by McNeil and Miller, respectively.

It seems reasonable to infer from the numbers in Table 3 that the lower bound (which is extremely simple to evaluate) is quite close, even for traffic intensities of two-thirds the saturation intensity, and it improves as the cycle time increases. For traffic intensities greater than this value, the upper bound given by Miller (which is also trivial to evaluate) is reasonably close, and again improves as the cycle time increases.

Of historic interest is a formula obtained by Webster[83] for the case of simple Poisson arrivals. Webster simulated the Poisson arrival stream and fitted a curve to the results obtained, the formula being

$$\frac{E[W]}{\lambda T^2} = \frac{r^2}{(1-\lambda s)} + \left(\frac{s}{T}\right)\frac{\lambda s}{2(1-r-\lambda s)(1-r)}$$
$$- \frac{0.65\{\lambda s/(1-r)\}^{5(1-r)+4/3}}{\{(1-r)T/s\}^{2/3}}, \quad (148)$$

where $r = R/T$. When λs is small, the right-hand side of Eq. 148 under-estimates the true value, the value for $r = \frac{1}{2}$, and $\lambda s = 0.165$ and $T/s = 20$ being 0.1727, compared with 0.1828 from Table 3. For values of λs close to $1 - r$, however, Webster's formula is extremely accurate; when $r = \frac{1}{2}$, $\lambda s = 0.495$, t gives 5.055 for $T/s = 20$ (against a true value of 5.104), 2.633 (against 2.640) for $T/s = 40$, and 1.429 (against 1.418) for $T/s = 80$.

In view of the apparent closeness of the crude lower bound for small to moderate traffic intensities and the crude upper bound for moderate to heavy flows, there seems little point in attempting to improve them. The upper and lower bounds for the overflow $E[Q(0)]$ are valid in general, but the expression (Eq. 127) for the mean delay is only valid for Darroch's arrival process and constant departure times or compound Poisson arrivals and independent, identically distributed departure times. The difficulties involved in an exact analysis for more general arrival and departure processes prompted Newell[74] to look for approximations, based on the representation of the traffic as a continuous fluid with stochastic properties. Using laws of large numbers, Newell obtained results insensitive to the detailed structure of the arrival and departure processes. We now briefly summarize Newell's work.

Considering first the case when the traffic intensity is sufficiently small for the overflow to be neglected, Newell used a heuristic graphical argument to conclude that for most reasonable arrival and departure processes $E[W]$ differs from Clayton's expression (Eq. 124) by an amount which is $O\{s(T - R)^{-1}\}$ as $s(T - R)^{-1} \to 0$. For flows which are so large that the overflow cannot be neglected, the following heuristic argument is used.

The expected total delay $E[W]$ is independent of the order in which the vehicles depart. Suppose, therefore, that vehicles depart according to the criterion "last come, first served." For this discipline the $Q(0)$ vehicles waiting at the commencement of the red period do not depart until any new arrivals have departed. For this type of departure process, however, the delay suffered by the new arrivals is the same as that which would exist if $Q(0) = 0$, and for large values of $(T - R)/s$, this is given approximately by Clayton's formula. If the traffic flow is sufficiently heavy, the original $Q(0)$ vehicles will now wait almost, if not all of, the cycle time T.

The expected delay per cycle for all vehicles is thus given approximately by

$$E[W] \sim \frac{\lambda R}{2(1 - \lambda s)}\left[R + \frac{2}{\lambda}\left(\frac{1 - \lambda s}{R/T}\right)E[Q(0)]\right]. \qquad (149)$$

This approximation for $E[W]$ may be compared with the exact expression for less general inputs given by Eq. 127. If the terms that are $O(s)$ are neglected, the expressions differ, but are identical if we put $1 - \lambda s = R/T$. This equality is never satisfied, since the condition for equilibrium is $1 - \lambda s < R/T$. However, the difference is small for the near-saturation situation, and this is the case which Newell considered. The advantage of Eq. 149 is that it holds for more general arrival and departure processes than does Eq. 127.

The next problem is to estimate the overflow. To do this Newell put

$$F_Q(x) = \text{Prob} \, [Q(0) \leq x], \tag{150}$$

and

$$F_{A-D}(x) = \text{Prob} \, [A - D \leq x], \tag{151}$$

where A and D are the numbers of arrivals and departures (if available), respectively, in a cycle. Assuming the difference between the traffic signal and the queue described by Eq. 140 is negligible, we have

$$Q(T) = \max \, (0, Q(0) + A - D). \tag{152}$$

Using Eqs. 150 and 151, Eq. 152 gives, in equilibrium,

$$F_Q(x) = \int_0^\infty F_Q(z) \, dF_{A-D}(x - z). \tag{153}$$

This is the Wiener–Hopf integral equation, which Lindley[84] used to describe the waiting time distribution for a simple queue. An explicit solution for $F_Q(z)$ is generally obtainable only if $F_{A-D}(x)$ has an exponential-type tail (Feller[70] discusses the reasons for this). This is not the case here, for the central limit theorem tells us that (for any reasonable arrival process) the distribution of $A - D$ is asymptotically normal as $s \to 0$. However, Newell obtained an approximation to $F_Q(x)$ by assuming that $E[Q(0)]$ is large compared with terms that are $O(\{T/s\}^{1/2})$. This is equivalent to saying that the expectation of $Q(0)$ is large compared with its standard deviation, in which case $Q(t)/E\{Q(0)\}$ can be regarded as a Brownian motion on a scale of time in which the cycle time itself is small. We then expand the $F_Q(x)$ in the integrand in the right-hand side of Eq. 153 in a Taylor series about $z = x$, and considering the first three terms only, obtain as a first approximation,

$$F_Q(x) \sim 1 - \exp \left(- \frac{2(1 - R/T - \lambda s)T/s}{\text{var} \, (A - D)} \, x \right),$$

whence

$$E[Q(0)] \sim \frac{\text{var} \, (A - D)}{2(1 - R/T - \lambda s)} \left(\frac{s}{T} \right). \tag{154}$$

The right-hand side of Eq. 154 is identical to the upper bound (Eq. 144) obtained by Miller[81] using a simple, elegant, and rigorous method.

Newell proceeded to obtain a closer approximation for $E[Q(0)]$ valid for light, moderate, and heavy flows, by assuming that the distribution of $A - D$

is precisely normal and solving Eq. 152 exactly. His result is

$$E[W] \sim \frac{\lambda R}{2(1 - \lambda s)} \left[R + \left(\frac{1 - \lambda s}{R/T} \right) \frac{(1 - R/T)I^*H(\mu)s}{2(1 - R/T - \lambda s)} \right], \qquad (155)$$

where

$$H(\mu) = \frac{2u^2}{\pi} \int_0^{\pi/2} \tan^2 \theta \{ \exp (\tfrac{1}{2}\mu^2 \cos^{-2}\theta) - 1 \}^{-1} \, d\theta$$

$$= 1 - (1.164 \cdots)\mu + \tfrac{1}{2}\mu^2 + \cdots \qquad \text{for} \qquad \mu \ll 1$$

$$= (2\pi)^{-1/2}\mu^{-2} \exp (-\tfrac{1}{2}\mu^2)\left(1 - \frac{3}{\mu^2} + \frac{3.5}{\mu^4} - \cdots + O(\exp(-\tfrac{1}{2}\mu^2)) \right)$$

$$\text{for} \qquad \mu \gg 1,$$

and

$$I^* = \lim_{s \to 0} \frac{\text{var} [A - D]}{(1 - R/T)T/s}.$$

This approximation is close when $1 - \lambda s$ is close to R/T and s is small. By comparing it with Webster's formula (Eq. 148) and using arguments of his own, Newell amended Eq. 155 to

$$E[W] \sim \frac{\lambda R}{2(1 - \lambda s)} \left[R + \left(\frac{1 - \lambda s}{R/T} \right) \frac{(1 - R/T)I^*H(\mu)s}{2(1 - R/T - \lambda s)} + \frac{s(1 - R/T)I^*}{(1 - \lambda s)} \right].$$

$$(156)$$

The main drawback in Eq. 155 arises from the fact that the heuristic analysis leading to Eq. 149 is only valid when λs is close to $1 - R/T$, if at all.

The Optimization Problem

Although many of the people who have studied the problem of evaluating $E[W]$ have motivated their work with the statement that the problem is of practical importance, very few have attempted the optimization problem. So far as we know, Miller[81] is the only author who has obtained analytic expressions for the optimum signal settings, in his case for a four-way intersection. Others have tackled the problem numerically for special input processes: Clayton[73] and Wardrop[85] for deterministic arrivals and Webster[83] for simple Poisson inputs. In addition, Gazis and Potts[86] have considered the problem of finding settings which minimize rush-hour delays, when the equilibrium condition is violated and the flows vary with time, and Grafton and Newell[87] have discussed an associated optimization problem.

For a reasonably general arrival process and constant departure times, Miller[81] obtained the formula

$$E[W] = \frac{\lambda R}{2(1 - \lambda s)}\left(RT + \frac{2}{\lambda}E[Q(0)] + s(I - 1 + \lambda s)\right), \qquad (157)$$

which is slightly different from the expression we have given in Eq. 122. The difference is due to the fact that Miller made an approximation in order to establish Eq. 157. In most situations $s/R \ll 1$, so the discrepancy is negligible. To find the optimum settings Miller used an approximation for $E[Q(0)]$. This approximation is based on the assumption that δ, defined in Eq. 140, satisfies

$$\frac{\text{var }[\delta]}{E[\delta]} \sim I.$$

Equation 154 can now be replaced by the approximate relation

$$E[Q(0)] \sim \frac{I\lambda s}{2(1 - R/T - \lambda s)} - \frac{I}{2}. \qquad (158)$$

Neglecting the term involving s, Eq. 157 becomes

$$E[W] \sim \frac{\lambda R}{2(1 - \lambda s)}\left(rT + \frac{(2\lambda s + R/T - 1)I}{\lambda(1 - R/T - \lambda s)}\right). \qquad (159)$$

Considering now a simple four-way junction, we number the directions 1, 2, 3, 4 in rotation and denote the flow parameters for direction i by λ_i, I_i, and s_i, respectively. The total lost time due to signal changes in one cycle is L, and the green period for the ith direction is G_i, where $G_1 = G_3$ and $G_2 = G_4$. Finally, define $\pi_i = G_i(T - L)^{-1}$ so that π_i is the fraction of effective green time available to the vehicles entering from direction i. The expression to be minimized is the average delay per motorist, and this is proportional to what Miller called the rate of delay D, where

$$D = \frac{1}{T}\sum_{i=1}^{4} E[W_i]. \qquad (160)$$

Using Eq. 159, we have

$$D \sim \sum_{i=1}^{4} \frac{\lambda_i[1 - \pi_i(1 - L/T)]}{1 - \lambda_i s_i}\left((1 - \pi_1)T + \pi_i L + \frac{[2\lambda_i s_i - \pi_i(1 - L/T)]}{\lambda_i[\pi_i(1 - L/T) - \lambda_1 s_1]}\right). \qquad (161)$$

We proceed with the optimization by first varying the cycle time T, keeping the allocation to each direction π_i fixed, and then varying the proportions π_i with T fixed. Neglecting terms that are $O(L/T)$, we deduce from Eq. 161

that

$$\frac{\partial D}{\partial T} \sim \sum_{i=1}^{4} \lambda_i \left(\frac{(1 - \pi_i)^2}{1 - \lambda_i s_i} - \frac{\pi_i s_i I_i L}{T^2 [\pi_i (1 - L/T) - \lambda_i s_i]} \right). \tag{162}$$

Similarly, by differentiating with respect to one of the proportions π_j and neglecting terms which are small, we have

$$\frac{\partial D}{\partial \pi_j} \sim -(T - L) \sum_{i=1}^{4} (-1)^{i-j} \lambda_i$$

$$\times \left(\frac{2[1 - \pi_i (1 - L/T)]}{1 - \lambda_i s_i} + \frac{s_i I_i}{T[\pi_i (1 - L/T) - \lambda_i s_i]^2} \right). \tag{163}$$

If we equate to zero the expressions on the right-hand sides of Eqs. 162 and 163, we obtain the optimal signal settings T and π_j.

A numerical study of the manner in which R varies with T for various values of the other parameters reveals that the rate of delay increases rapidly for cycle lengths less than the optimum, but only slowly if T is too long. To err on the safe side, optimum settings are found for the direction of traffic with the largest value of the traffic intensity λs. Then from Eq. 162, the cycle time setting satisfies

$$\frac{(1 - \pi_i)^2}{1 - \lambda_i s_i} = \frac{\pi_i s_i I_i L}{T^2 [\pi_i (1 - L/T) - \lambda_i s_i]^2}.$$

Extracting T, we obtain

$$T = \frac{L + [s_i I_i L(1 - \lambda_i s_i)/\pi_i]^{1/2}(1 - \pi_i)^{-1}}{1 - \lambda_i s_i / \pi_i}. \tag{164}$$

For this value of T, Miller found, by inspection of the right-hand side of Eq. 163, that the first terms in this expression are small, and so for each pair of competing directions (i, j) approximate ratio settings are derived by solving

$$\frac{\lambda_i s_i I_i}{[\pi_i (1 - L/T) - \lambda_i s_i]^2} = \frac{\lambda_j s_j I_j}{[\pi_j (1 - L/T) - \lambda_j s_j]^2}.$$

Putting $\pi_j = 1 - \pi_i$ and rearranging, this gives

$$\pi_i = \frac{(\lambda_i s_i I_i)^{1/2} + (1 - L/T)^{-1}(\lambda_i s_i \lambda_j s_j)^{1/2}[(\lambda_i s_i I_i)^{1/2} - (\lambda_j s_j I_i)^{1/2}]}{(\lambda_i s_i I_i)^{1/2} + (\lambda_j s_j I_j)^{1/2}}. \tag{165}$$

Clearly Eq. 165 cannot be satisfied by both of the values of j associated with any particular choice of i, so it is assumed that j is the direction opposing i for which $\lambda_j s_j$ is greatest. The optimum settings are now given by Eqs. 164

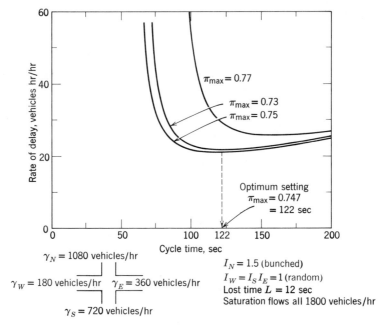

FIGURE 2. Delays resulting from various signal settings.

and 165. These formulas are much easier to use if, as Miller suggested, the term $(1 - L/T)^{-1}$ is replaced by a constant, say 1.2; this means that π_i is given directly by Eq. 165, and the optimum cycle time is given by substituting this value into Eq. 164.

Figure 2 shows the rate of delay as calculated from Eq. 159 plotted against cycle time for the illustrated four-way junction. At the calculated settings, the traffic intensity is 0.89. $\pi_{\max}$ is the larger of the two proportions π.

The Vehicle-Actuated Traffic Signal

The type of signal considered so far has green and red periods of fixed length, and there occur from time to time periods during which traffic is blocked in one direction even though there is no traffic in the other direction(s). In situations like this it would clearly be better to change the signal as soon as the favored queue empties, instead of forcing the stopped traffic to wait until the end of the cycle. In this case the traffic signal is said to be vehicle-actuated and the durations of the red and green periods vary, depending on the traffic flow in each direction, whereas for the fixed-cycle traffic signal there is no such dependence.

If we assume, for the moment, that the departure processes for the various streams are identical, then (since there is inevitably a delay associated with each signal change) the best policy for reducing the expected delay is to change the signal as soon as, but not before, the favored queue has emptied. In this case the only problem is to decide exactly when a queue has emptied, that is, to choose the maximum allowable headway between vehicles approaching the intersection. In practice this will determine the positions of the roadway detectors which activate the signals.

For practical purposes, however, vehicle-actuated signals have a *maximum* green period, so that the signal changes after a certain time even if the favored queue is not empty. The reason for this is to ensure that an individual motorist does not face a continuous red light for more than a fixed time; howevei, the *expected* delay is increased by such a policy.

In the case of the fixed-cycle signal it was possible to determine the expected delay, and thus the optimum signal settings, by considering each traffic stream separately, since there is no relation between the streams. For the vehicle-actuated signal this is no longer possible, and the analysis leading to expressions for the delay becomes increasingly complicated as the number of interacting streams increases. Accordingly, let us consider first the simplest situation, namely, the intersection of two one-way streets. Even in this case the analysis is complicated unless it is assumed that there is no upper limit to the duration of a green period (see Lehoczky[88]). This model was considered first by Tanner[54] and more recently by Darroch, Newell, and Morris.[67] In both papers it is assumed that arrivals are (simple) Poisson, but Darroch et al. allow for variable departure times and lost times while Tanner does not. Tanner's model is more realistic in dealing with the change-over phenomenom from green to one stream to another, but this causes analytic difficulties that do not arise in the other model. The differences are spelled out in the paper by Darroch et al.

In the analysis that follows, we simplify matters by assuming that all the departure times are of fixed duration s, that the lost times are of fixed length L, and that the arrival streams are simple Poisson with rate λ_i for direction i. Finally, it is assumed that the traffic signal controls the two intersecting traffic streams in such a way that for each direction the signal is green until any existing queue is discharged and a headway of at least β_i is detected in the subsequent arrivals. This is a special case of the model of Darroch et al., but the generalization to independent and identically distributed departure times and lost times (with different distributions for each stream) is straightforward. The problem is to find the values of β_1 and β_2 that minimize the total rate of delay at the intersection.

Let the subscript i refer to direction (or stream) i. Since there is no overflow at the end of a given period, the expected total wait per cycle is given by

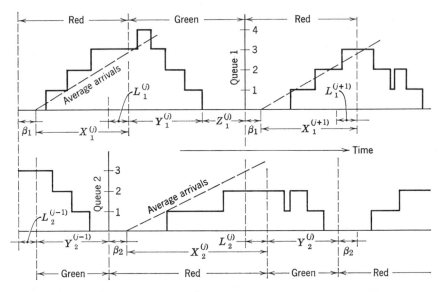

FIGURE 3. A realization of the traffic queue at vehicle-actuated traffic signals.

Eq. 120 without the term involving $E[Q(0)]$. The red period is now a random variable, so we have, in equilibrium,

$$E[W_i] = \frac{\lambda_i}{2(1 - \lambda_i s)} \left\{ E[R_i^2] + sE[R_i] \left(1 + \frac{1}{1 - \lambda_i s} \right) \right\}. \qquad (166)$$

To evaluate the first two moments of R_i we consider two successive cycles and obtain the relations (see Figure 3)

$$R_1^{(j+1)} = 2L + Y_2^{(j)} + Z_2^{(j)} - \beta_1,$$
$$R_2^{(j)} = 2L + Y_1^{(j)} + Z_1^{(j)} - \beta_2, \qquad (167)$$

where Y_i is the time for the queue to be discharged from stream i, Z_i is the additional time that elapses before the beginning of a headway of at least β_i, and the superscript j denotes the cycle number. Clearly, we must impose the conditions $\beta_i \leq 2L$; otherwise, the right-hand sides of Eq. 167 could be negative.

Suppose there are Q_i vehicles queueing in direction i when the signal changes from red to green during a cycle. Then $Y_i^{(j)}$ is the convolution of $Q_i^{(j)}$ busy periods in a queue with simple Poisson arrivals (with rate λ_i) and constant service times (of duration s). Using known results from queueing theory (see, for example, Cox and Smith,[1] p. 149) we have

$$E[Y_i \mid Q_i] = \frac{Q_i s}{1 - \lambda_i s},$$

$$E[Y_i^2 \mid Q_i] = E^2[Y_i \mid Q_i] + \frac{Q_i \lambda_i s^3}{(1 - \lambda_i s)^3}.$$

But $Q_i^{(j)}$ is the number of arrivals in the red period $R_i^{(j)}$, and is thus Poisson-distributed with mean $\lambda_i R_i^{(j)}$. Consequently, taking expectations in the above we get

$$E[Y_i] = \frac{\lambda_i s E[R_i]}{1 - \lambda_i s},$$

$$E[Y_i^2] = \lambda_i s^2 \frac{\{E[R_i] + \lambda_i(1 - \lambda_i s)E[R_i^2]\}}{(1 - \lambda_i s)^3}. \tag{168}$$

Taking expected values in Eq. 167 and using Eq. 168, we find

$$E[R_1] = 2L - \beta_1 + E[Z_2] + \frac{\lambda_2 s E[R_2]}{1 - \lambda_2 s},$$

$$E[R_2] = 2L - \beta_2 + E[Z_1] + \frac{\lambda_1 s E[R_1]}{1 - \lambda_1 s},$$

whence

$$E[R_1] = \frac{(1 - \lambda_1 s)\{2L - \beta_1 + \lambda_2 s(\beta_1 - \beta_2) + \lambda_2 s E[Z_1] + (1 - \lambda_2 s)E[Z_2]\}}{1 - (\lambda_1 + \lambda_2)s}. \tag{169}$$

The expression for $E[R_2]$ is obtained simply by interchanging the subscripts 1 and 2 in Eq. 169, for reasons of symmetry.

We can now obtain the average cycle time. It is, using Figure 3,

$$E[T] = E[R_1] + E[R_2] - 2L + \beta_1 + \beta_2. \tag{170}$$

Equations 169 and 170 yield

$$E[T] = \frac{2L + (1 - \lambda_1 s)E[Z_1] + (1 - \lambda_2 s)E[Z_2] - \beta_1 \lambda_1 s - \beta_2 \lambda_2 s}{1 - (\lambda_1 + \lambda_2)s}. \tag{171}$$

To obtain $E[R_i^2]$, we square both sides of Eqs. 167 and take expectations. Since Y_i and Z_i are independent, we obtain

$$E[R_1^2] = (2L - \beta_1)^2 + 2(2L - \beta_1)\{E[Y_2] + E[Z_2]\}$$
$$+ E[Y_2^2] + 2E[Y_2]E[Z_2] + E[Z_2^2], \tag{172}$$

together with a similar expression for $E[R_2^2]$. Explicit formulas are now obtainable using Eqs. 168 and 169, but they are quite complicated.

To obtain the total rate of delay, that is, $\{E[W_1] + E[W_2]\}/E[T]$, it only remains to determine the first two moments of Z_i. Now Z_i is just the waiting time until the beginning of the first gap of at least β_i in a simple Poisson stream with parameter λ_i, so from Eq. 28

$$E[Z_i + \beta_i] = \frac{\exp(\lambda_i\beta_i) - 1}{\lambda_i},$$

$$E[Z_i + \beta_i]^2 = 2\exp(\lambda_i\beta_i)\frac{\exp(\lambda_i\beta_i) - 1 - \lambda_i\beta_i}{\lambda_i^2}.$$

(173)

It is easily shown that the condition for equilibrium is

$$(\lambda_1 + \lambda_2)s < 1,$$ (174)

as we would expect.

It is now possible to determine the values of β_1 and β_2 that minimize the total delay per unit time: $\{E[W_1] + E[W_2]\}/E[T]$, say, but graphical or numerical methods are required except in special cases. The formula for the rate of delay simplifies considerably if the flow is nearly saturated, that is, if $(\lambda_1 + \lambda_2)s$ is close to unity. In this case the right-hand side of Eq. 169 is large (since its denominator is small), and thus, if we use Eq. 168, $E[Y_2]$ and $E[Y_2^2]$ are the dominant terms in the right-hand side of Eq. 172. Neglecting the remaining terms and using Eq. 168, we get

$$E[R_1^2] \approx \frac{\lambda_2^2s^2}{(1 - \lambda_2s)^2}E[R_2^2] + \left(\frac{2(2L - R_1 + E[Z_2])}{1 - \lambda_2s} + \frac{\lambda_2s^2}{(1 - \lambda_2s)^3}\right)E[R_2],$$

(175)

together with a similar expression for $E[R_2^2]$. Also, when $(\lambda_1 + \lambda_2)s$ is close to unity, Eqs. 169 and 171 yield

$$E[R_1] \approx \lambda_2sE[T], \qquad E[R_2] \approx \lambda_1sE[T].$$ (176)

Now solving Eq. 175 as simultaneous equations and using Eqs. 176, we get

$$\lambda_i^2E[R_i^2] \approx \frac{\lambda_1\lambda_2s^2\{(\lambda_1 + \lambda_2)(1 + 4\lambda_1\lambda_2Ls) + \\ + 2\lambda_1\lambda_2s(\lambda_1E[Z_2] - \lambda_1\beta_1 + \lambda_2E[Z_1] + \lambda_2\beta_2)\}E[T]}{2\{1 - (\lambda_1 + \lambda_2)s\}}.$$

If we use Eq. 171, this reduces further to

$$\lambda_i^2E[R_i^2] \approx \lambda_1\lambda_2E[T]\left(E[T] + \frac{s - 4L(1 - \lambda_1\lambda_2s^2)}{2\{1 - (\lambda_1 + \lambda_2)s\}}\right).$$ (177)

Equations 166 and 177 yield for the limiting rate of delay, as $(\lambda_1 + \lambda_2)s$ tends to unity

$$\frac{E[W_1] + E[W_2]}{E[T]} \approx \frac{E[T]}{s} + \frac{1 - 4(1 - \lambda_1\lambda_2s^2)L/s}{2[1 - (\lambda_1 + \lambda_2)s]}. \tag{178}$$

The right-hand side of Eq. 178 only involves β_1 and β_2 through the mean cycle time $E[T]$. The values of β_1 and β_2 for which $E[T]$ is a minimum occur when $\partial E[T]/\partial\beta_i = 0$. Inserting Eqs. 173 in Eq. 171 and differentiating, the optimum values are found to be

$$\beta_i = \frac{1}{\lambda_i}\log\left(\frac{1}{1 - \lambda_i s}\right) \qquad i = 1, 2. \tag{179}$$

The requirement $\beta_i \leq 2L$ must also be satisfied, and Eq. 179 gives the values of β_1 and β_2 that minimize the total rate of delay for high flows. The usefulness of this result is reduced by the fact that in real traffic situations, as stated earlier, there are upper limits to the red periods, and these limits may be taken as infinity only if the traffic is sufficiently light.

Another special case for which the formulas simplify considerably is the symmetric intersection, in which $\lambda_1 = \lambda_2 = \lambda$. In this case it may be shown, using Eqs. 166, 169, 171, and 172, that

$$\frac{E[W_1] + E[W_2]}{E[T]} = \frac{\lambda(\frac{1}{2}E[T] + L - \beta)^2}{(1 - \lambda s)E[T]}$$
$$+ \frac{2\lambda s(\frac{1}{2}E[T] + L - \beta) + (1 - \lambda s)(e^{2\lambda\beta} - \lambda\beta e^{\lambda\beta} - 1)/\lambda}{(1 - 2\lambda s)E[T]}. \tag{180}$$

The optimal value of β is now obtained by differentiating and equating to zero the right-hand side of Eq. 180, but it does not seem possible to obtain an explicit value. Darroch et al. conjecture that the value of β which minimizes the total rate of delay is always somewhat higher than the value which minimizes the average cycle time. From Eq. 171, $E[T]$ is minimized in the symmetric case when $\beta = -\lambda^{-1}\log(1 - \lambda s)$, which is, of course, Eq. 179 for this case.

In a later paper Dunne[89] considers a model which he claims to be the discrete analog to that of Darroch et al. This claim is hardly justified, however, since the departure times and lost times are assumed constant (these constants being the same for each direction), and there is no allowance for the extension times Z_i. The only facet of Dunne's model which does not reduce to a special case of that of Darroch et al. is the arrival process, which is assumed to be binomial with the h intervals defining it coincident with the constant departure times. There is, however, one advantage in considering this model—as would be expected, the expressions for the average delay are much simpler than those given by Eqs. 166, 169, and 172.

The analog to Eq. 166 is simply

$$E[W_i] = \frac{\lambda_i}{2(1 - \lambda_i s_i)} \{E[R_i^2] + 2sE[R_i]\}, \tag{181}$$

which, like Eq. 166, may be obtained by putting $E[Q(0)] = 0$ in Eq. 122, together with the other specializations for constant departure times and binomial arrivals. The Laplace–Stieltjes transform of the R_i for this model was previously obtained by Dunne and Potts.[90]

Darroch et al. state that the main drawback of their model is the restriction to simple Poisson arrivals. The analysis can be carried through quite simply for a somewhat more general arrival process, namely, the arrival process first obtained by Darroch[78] and discussed above on p. 143, if the departure times are lost times or fixed multiples of h, where h is the time unit for the arrival process. This model, studied by Lehoczky,[88] has the advantage of allowing for any index of dispersion in the arrival process. If an index of dispersion not less than unity suffices, it is straightforward to generalize the results of Darroch, Newell, and Morris to compound Poisson arrivals. In the concluding paragraphs the paper is described as the obvious second step to earlier work involving continuous approximations by Grace, Morris, and Pak-Poy.[91] It is stated that the third step would be to eliminate the restriction to simple Poisson arrivals and allow for a more general stationary arrival process. Newell[68] has considered the vehicle-actuated traffic signal with general stationary arrival and departure processes by using diffusion theory approximations. Among other things, he shows that the cycle time has approximately a gamma distribution and that in certain circumstances the average delay per car is less than that for fixed-cycle signals by about a factor of 3.

For intersections of one-way streets, the mean delay for vehicle-actuated signals is explicitly obtainable for any number of intersecting streams, but the algebra is formidable even in the case of a three-way intersection (see Hoshi[92]). It is difficult to carry out an exact analysis for two-way streets; in this case one must decide whether to change the signal as soon as one of the opposing queues has been discharged, or to wait until both have emptied. In the former (latter) case the green period is the minimum (maximum) of two random variables, and thus the moments are not easily found.

For an intersection of one-way streets controlled by vehicle-actuated signals a simple lower bound to the expected delay per vehicle for the model of Darroch et al. may be found as follows. If the lost times are assumed zero and there are no extension times then the delay is reduced. Since there is no loss due to change-over, the system behaves just like a queue with known service time distribution and an input process which is the superposition of the two arrival processes. The average waiting time per vehicle in the case of simple Poisson arrivals is thus given by Pollaczek's formula (see, for example, Cox and Smith,[1] p. 55) and consequently

$$E[W] \leq s\left(1 + \frac{(\lambda_1 + \lambda_2)s(1 + C^2)}{2(1 - \lambda_1 s - \lambda_2 s)}\right), \qquad (182)$$

where λ_i is the arrival rate for direction i, s is the mean, and C the coefficient of variation of the departure times.

Garwood's Model

In the situation described in the preceding section there are two streams of traffic with neither having priority. We conclude with discussion of a model in which one stream has priority over the other, and which may be described as follows.

A main stream of traffic intersects a subsidiary stream. The signal is green for main stream vehicles until a subsidiary stream vehicle arrives. If no main stream vehicles arrive within the time τ, where τ is fixed, then the signal changes to allow the subsidiary stream vehicle to cross the intersection. Otherwise the signal changes as soon there is a headway of at least τ in the main stream traffic. When the signal changes, the main stream vehicle, together with any others that have formed a queue, departs. When these vehicles have departed, the signal switches back to green for the main stream traffic. Clearly, the headway detectors for the main stream vehicles will normally be placed at some distance back from the intersection. This is to ensure that the headway is used by the minor stream vehicle. If the distance is too great, however, the main stream vehicle will not have time to cross the intersection.

If the traffic in the main stream is sufficiently heavy, a subsidiary stream motorist may have to wait a long time before being permitted to depart, and normally an upper limit, T, where T is fixed, is imposed on this waiting time. In this case, if all the headways in the main stream are smaller than τ, the signal automatically changes at time T after the arrival of the minor stream vehicle.

The model described above was analyzed in detail by Garwood[45] by means of elegant combinatorial arguments, but it will be recognized that the results can be obtained directly from Eqs. 26 and 27.

References

1. D. R. Cox and W. L. Smith, *Queues*, London: Methuen, 1961.
2. D. R. Cox, *Renewal Theory*, London: Methuen, 1962.
3. W. F. Adams, "Road Traffic Considered as a Random Series," *J. Inst. Civil. Engrs.*, **4**, 121–130 (1936).

4. G. H. Weiss and R. Herman, "Statistical Properties of Low Density Traffic," *Quart. Appl. Math.*, **20,** 121–130 (1962).

5. L. Breiman, "The Poisson Tendency in Traffic Distribution," *Ann. Math. Statist* , **34,** 308–311 (1963)

6. T. Thedeen, "A Note on the Poisson Tendency in Traffic," *Ann. Math. Statist.*, **35,** 1823–1824 (1964).

7. A. Schuhl, "The Probability Theory Applied to Distributions to Vehicles on Two-Lane Highways," *Poisson and Traffic*, New Haven: Eno Foundation, 1955, pp. 59–72.

8. D. J. Buckley, "Road Traffic Distributions," *Proc. Aust. Road Res. Board*, **1,** 153–187 (1962).

9. D. J. Buckley, "A Semi-Poisson Model of Traffic Flow, *Transport Sci.*, **2,** 107–133 (1968).

10. A. J. Miller, "A Queueing Model for Road Traffic Flow," *Proc. Roy. Statist. Soc.*, **B-23,** 64–90 (1961).

11. H. E. Daniels, "Mixtures of Geometric Distributions," *Proc. Roy. Statist. Soc.* **B-23,** 409–413 (1961).

12. R. M. Oliver, "Distribution of Gaps and Blocks in a Traffic Stream," *Operations Res.*, **10,** 197–217 (1962).

13. M. J. Beckmann, C. B. McGuire, and C. B. Winsten, *Studies in the Economics of Transportation*, New Haven: Yale University Press, 1956.

14. R. L. Moore, "Psychological Factors of Importance in Traffic Engineering," *Intern. Study Week in Traffic Engineering* (Stresa, Italy', London: World Touring and Automobile Association, 1956.

15. W. L. Gibbs, "Driver Gap Acceptance at Intersections," *J. Appl. Psych.*, **52,** 200–211 (1968).

16. P. M. Hurst, K. Perchonok, and E. L. Seguin, "Vehicle Kinematics and Gap Acceptance," *J. Appl. Psych.*, **52,** 321–324 (1968).

17. M. S. Raff, "The Distribution of Blocks in an Uncongested Stream of Automobile Traffic," *J. Amer. Statist. Assoc.*, **46,** 114–123 (1951).

18. J. C. Tanner, "The Delay to Pedestrians Crossing a Road," *Biometrika*, **38,** 383–392 (1951).

19. R. Herman and G. H. Weiss, "Comments on the Highway Crossing Problem," *Operations Res.*, **9,** 828–840 (1961).

20. D. R. McNeil and J. H. T. Morgan, "Estimating Minimum Gap Acceptances for Merging Motorists," *Transport Sci.*, **2,** 265–277 (1968).

21. W. R. Blunden, C. M. Clissold, and R. B. Fisher, "Distribution of Acceptance Gaps for Crossing and Turning Maneuvers," *Proc. Aust. Road Res. Board*, **1,** 188–205 (1962).

22. D. R. Drew, L. R. LaMotte, J. H. Buhr, and J. A. Wattleworth, "Gap Acceptance in the Freeway Merging Process," Report 430–2, Texas Transportation Institute, 1967.

23. D. R. Drew, J. H. Buhr, and R. H. Whitson, "The Determination of Merging Capacity and Its Application to Freeway Design and Control," Report 430–4, Texas Transportation Institute, 1967.

24. J. A. Wattleworth, J. H. Buhr, D. R. Drew, and F. A. Gerig, Jr., "Organizational Effects of Some Ramp Geometrics on Freeway Merging," Report 430-3, Texas Transportation Institute, 1967.

25. P. Solberg and J. D. Oppenlander, "Lag and Gap Acceptances at a Stop-Controlled Intersection," *Highway Res. Rec.*, **118**, 48–67 (1966).

26. F. A. Wagner, "An Evaluation of Fundamental Driver Decisions and Reactions at an Intersection," *Highway Res. Rec.*, **118**, 68–84 (1966).

27. R. D. Worrall, D. W. Coutts, H. Echterhoff-Hammerschmid, and D. S. Berry, "Merging Behavior at Freeway Entrance Ramps: Some Elementary Empirical Considerations," *Highway Res. Rec.*, **157**, 108–143 (1967).

28. J. Cohen, E. J. Dearnaley, and C. E. M. Hansel, "The Risk Taken Crossing a Road," *Operations Res. Quart.*, **6**, 120–128 (1955).

29. A. J. Miller, "Nine Estimators of Gap-Acceptance Parameters," *Proc. 5th Intern. Symp. on the Theory of Traffic Flow and Transportation*, G. F. Newell, Ed. New York, Elsevier, 1972, pp. 215–235.

30. R. Ashworth, "A Note on the Selection of Gap Acceptance Criteria for Traffic Simulation Studies," *Transport Res.*, **2**, 171–175 (1968).

31. R. Ashworth, "The Analysis and Interpretation of Gap Acceptance Data," *Transport Sci.*, **4**, 270–280 (1970).

32. D. H. Evans, R. Herman, and G. H. Weiss, "The Highway Merging and Queueing Problem," *Operations Res.*, **12**, 832–857 (1964).

33. G. F. Yeo and B. Weesakul, "Delays to Road Traffic at an Intersection," *J. Appl. Prob.*, **2**, 297–310 (1964).

34. A. G. Hawkes, "Gap Acceptance in Road Traffic," *J. Appl. Prob.*, **5**, 84–92 (1968).

35. G. H. Weiss, "Effects of a Distribution of Gap Acceptance Functions on Pedestrian Queues," *J. Res. Natl. Bur. Stds.*, **68B**, 13–15 (1964).

36. G. H. Weiss, "The Intersection Delay Problem with Correlated Gap Acceptance," *Operations Res.*, **14**, 614–619 (1966).

37. G. H. Weiss, "The Intersection Delay Problem with Gap Acceptance Function Depending on Space and Time," *Transport Res.*, **1**, 367–371 (1967).

38. D. E. Blumenfeld and G. H. Weiss, "On the Robustness of Certain Assumptions in the Merging Delay Problem," *Transport Res.*, **4**, 125–139 (1970).

39. D. E. Blumenfeld and G. H. Weiss, "On Queue Splitting to Reduce Waiting Times," *Transport Res.*, **4**, 141–144 (1970).

40. G. H. Weiss and A. A. Maradudin, "Some Problems in Traffic Flow," *Operations Res.*, **10**, 74–104 (1962).

41. A. J. Mayne, "Some Further Results in the Theory of Pedestrians and Road Traffic," *Biometrika*, **41**, 375–389 (1954).

42. D. R. McNeil and J. T. Smith, "A Comparison of Motorist Delays for Different Merging Strategies," *Transport. Sci.*, **3**, 239–254 (1969).

43. G. H. Weiss, "The Intersection Delay Problem with Mixed Cars and Trucks," *Transport Res.*, **3**, 195–199 (1969).

44. D. C. Gazis, G. F. Newell, P. Warren, and G. H. Weiss, "The Delay Problem for Crossing an *n* Lane Highway," *Proc. 3rd Intern. Symp. on Traffic Flow*, New York, Elsevier, 1967, pp. 267–279.

45. F. Garwood, "An Application of the Theory of Probability in the Operation of Vehicular Controlled Traffic Signals," *J. Roy. Statist. Soc. Suppl.*, **7**, 65–77 (1940).

46. F. A. Haight, E. F. Bisbee, and C. Wojcik, "Some Mathematical Aspects of the Problem of Merging," *Highway Res. Board Bull.*, **356** (1962).

47. H. Mine and T. Mimura, "Highway Merging Problem with Acceleration Lane," *Transport Sci.*, **3**, 205–213 (1969).

48. D. E. Blumenfeld and G. H. Weiss, "Merging from an Acceleration Lane," *Transport Sci.*, **5**, 161–168 (1971).

49. G. H. Weiss, "An Analysis of Pedestrian Queueing," *J. Res. Natl. Bur. Stds.*, **67B**, 229–243 (1963).

50. J. D. C. Little, "Approximate Expected Delays for Several Maneuvers by a Driver in Poisson Traffic," *Operations Res.*, **9**, 39–52 (1961).

51. R. J. Smeed, "Theoretical Studies and Operational Research on Traffic and Traffic Congestion," *Bull. Intern. Statist. Inst.*, **36**, 347–375 (1958).

52. R. J. Smeed, "Aspects of Pedestrian Safety," *J. Transport Econ. Policy*, **2**, No. 3, 1–25 (1968).

53. J. G. Wardrop, "Journey Speed and Flow in Central Urban Areas," *Traffic Engr. Cont.*, **9**, 528–532 (1968).

54. J. C. Tanner, "A Problem of Interference Between Two Queues," *Biometrika*, **40**, 58–69 (1953).

55. A. G. Hawkes, "Queueing at Traffic Intersections," *Proc. 2nd Intern. Symp. on the Theory of Traffic Flow*, edited by J. Almond, Paris: Office of Economic Cooperation and Development, 1965, pp. 190–199.

56. A. G. Hawkes, "Queueing for Gaps in Traffic," *Biometrika*, **52**, 79–85 (1965).

57. L. Takacs, "Investigation of Waiting Time Problems by Reduction to Markov Processes," *Acta Math. Hung.*, **6**, 101–129 (1955).

58. J. C. Tanner, "A Theoretical Analysis of Delays at an Uncontrolled Intersection," *Biometrika*, **49**, 163–170 (1962).

59. D. Gaver, "Accommodation of Second-Class Traffic," *Operations Res.*, **11**, 72–87 (1963).

60. D. J. Buckley and W. R. Blunden, "Some Delay-Flow Characteristics for Conflicting Traffic Streams," Ref. 55, pp. 167–181.

61. G. F. Yeo, "Single-Server Queues with Modified Service Mechanisms," *J. Aust. Math. Soc.*, **2**, 499–502 (1962).

62. A. G. Hawkes, "Delay at Traffic Intersections," *J. Roy. Statist. Soc.*, **B**, 202–212 (1966).

63. G. F. Newell, "The Effects of Left Turns on the Capacity of a Road Intersection," *Quart. Appl. Math.*, **17**, 67–76 (1959).

64. D. H. Reid, "A Mathematical Model for Delays Caused by Right-Turning Vehicles at an Uncontrolled Intersection," *J. Appl. Prob.*, **4**, 180–191 (1967); "Delays Caused by Right-Turning Vehicles," *Transport Sci.*, **2**, 160–171 (1968).

65. R. M. Oliver and E. F. Bisbee, "Queueing for Gaps in High Flow Traffic Streams," *Operations Res.*, **10**, 105–114 (1962).

66. A. J. Miller, "Settings for Fixed-Cycle Traffic Signals," *Proc. 2nd Conf. Aust. Road Res. Board*, **1**, 342–365 (1964).

67. J. N. Darroch, G. F. Newell, and R. W. J. Morris, "Queues for a Vehicle-Actuated Traffic Light," *Operations Res.*, **12**, 882–895 (1964).

68. G. F. Newell, "Properties of Vehicle Actuated Signals: I. One-Way Streets," *Transport Sci.*, **3**, 30–52 (1969).

69. J. D. Gordon and A. J. Miller, "Right-Turn Movements at Signalized Intersections," *Proc. Third Conf. Aust. Road Res. Board*, **1**, 446–459 (1966).

70. W. Feller, *An Introduction to Probability Theory and its Applications*, New York: Wiley, 1966, Vol. II.

71. D. R. McNeil, "A Solution to the Fixed Cycle Traffic Light Problem for Compound Poisson Arrivals," *J. Appl. Prob.*, **5**, 624–635 (1968).

72. D. J. Daley and D. R. Jacobs, "The Total Waiting Time in a Busy Period of a Stable Single-Server Queue, II," *J. Appl. Prob.*, **6**, 573–583 (1969).

73. A. J. H. Clayton, "Road Traffic Calculations," *J. Inst. Civil. Engrs.*, **16**, No. 7, 247–284 (1941); **16**, No. 8, 588–594 (1941).

74. G. F. Newell, "Approximation Methods for Queues with Application to the Fixed-Cycle Traffic Light, *SIAM Rev.*, **7**, 223–240 (1965).

75. D. J. Daley, "The Total Waiting Time in a Busy Period of a Stable Single-Server Queue, I.," *J. Appl. Prob.*, **6**, 565–572 (1969).

76. F. A. Haight, "Overflow at a Traffic Light," *Biometrika*, **46**, Nos. 3 and 4, 420–424 (1959).

77. G. F. Newell, "Queues for a Fixed-Cycle Traffic Light," *Ann. Math. Statist.*, **31**, 589–597 (1960).

78. J. N. Darroch, "On the Traffic-Light Queue," *Ann. Math. Statist.*, **35**, 380–388 (1964).

79. C. D. Crommelin, "Delay Probability Formulae When the Holding Times Are Constant," *Post Office Elect. Engr. J.*, **25**, 41–50 (1932).

80. D. C. Kleinecke, "Discrete Time Queues at a Periodic Traffic Light," *Operations Res.*, **12**, 809–814 (1964).

81. A. J. Miller, "Setting for Fixed-Cycle Traffic Signals," *Operations Res.*, **14**, 373–386 (1963).

82. V. Siskind, "The Fixed-Cycle Traffic Light Problem: A Note on a Paper by McNeil," *J. Appl. Prob.*, **7**, 245–248 (1970).

83. F. V. Webster, "Traffic Signal Settings," *Road. Res. Tech. Paper No. 39*, British Road Res. Lab., 1958.

84. D. V. Lindley, "The Theory of Queues with a Single Server," *Proc. Camb. Phil. Soc.*, **48**, 277–289 (1952).

85. J. G. Wardrop, "Some Theoretical Aspects of Road Traffic Research," *Proc. Inst. Civil Engrs. Pt. II*, **1**, 325–378 (1952).

86. D. C. Gazis and R. B. Potts, "The Over-Saturated Intersection," Ref. 55.

87. R. B. Grafton and G. F. Newell, "Optimal Policies for the Control of an Under-Saturated Intersection," Ref. 55, pp. 239–257.

88. J. Lehoczky, "The Variable Cycle Traffic Light," Ph.D. dissertation, Stanford University, 1969.

89. M. C. Dunne, "Traffic Delay at a Signalized Intersection with Binomial Arrivals," *Transport Sci.*, **1**, 24–31 (1967).

90. M. C. Dunne and R. B. Potts, "Analysis of a Computer Control of an Isolated Intersection," Ref. 44, pp. 258–266.

91. M. G. Grace, R. W. J. Morris, and P. G. Pak-Poy, "Some Aspects of Intersection Capacity and Traffic Signal Control by Computer Simulation," *Proc. Second Conf. Aust. Road Res. Board*, **1**, 274–304 (1964).

92. T. Hoshi, "Some Problems Involving Delays at Traffic Intersections," Ph.D. dissertation, The Johns Hopkins University, 1969.

93. W. L. Stevens, "Solution to a Geometrical Problem in Probability," *Ann. Eugen.*, **9**, 315–320 (1939).

94. T. Thedeen, "Delays at Pedestrian Crossings of the Push-Button Type," *Proc. 4th Symp. on Theory of Traffic Flow*, Bonn: Bundesminister für Verker, 1969, pp. 127–130.

CHAPTER 3

Traffic Control—Theory and Application

Denos C. Gazis

++

Contents

I. Introduction		175
II. Objectives and Principles of Operation of Traffic Control Systems		178
III. Synchronization Schemes		180
IV. Heuristic Control Algorithms for Microcontrol		199
	Single Intersections 199	
	Systems of Intersections 204	
V. Oversaturated Systems		209
	Single Intersections 209	
	Complex Oversaturated Systems 216	
	A Pair of Intersections 216	
	Spill-Back from an Exit Ramp of an Expressway 218	
	Reversible Lanes 221	
	Generalization to Complex Congested Transportation Systems 224	
VI. Control of Critical Traffic Links		229
	Control of Expressways 231	
	Control of Traffic at the Lincoln Tunnel 233	
	Measurement of Densities 234	
	References	237

I. INTRODUCTION

"Ladies and Gentlemen will order their Coachmen to take up and set down with their Horse Heads to the East River, to avoid confusion." Thus, according to Kane's *Famous First Facts*,[1] stated a New York City ordinance of 1791 establishing the first one-way street regulation. The regulation was incidental to performance at the John Street Theater, and there is no record indicating how well it worked. It is fair to say, however, that Confusion in New York City has managed not only to survive such assaults but also to expand considerably in scope. Things have changed, of course, over the last 180 years. We have abandoned horses and moved to a much slower but safer mode of transportation, the car and its ubiquitous companion, the double-parked truck. Confusion reigns eternal, only now it is controlled not only by one-way regulations but also by traffic lights. The lights, together with the basic instinct of self-preservation, by and large guarantee that no two cars occupy the same space at the same time. It would be tempting to attribute the above accomplishments to Yankee ingenuity, except that things are about the same in Tokyo and in London, only the British would never capitalize Confusion.

The above preamble is simply meant to reiterate something we all know: traffic conditions are bad in the cities and they do not appear to be getting any better. But some progress is being made, although it may not always be noticeable to the motorist struggling through the rush hour. Over the last decade, in particular, many contributions have been made to the study of traffic and the development of a methodology for its improvement through appropriate control. In this chapter we shall discuss some of this work with an emphasis on the foundations of a methodology for the control of traffic. Before this discussion, a few remarks about the evolution of traffic control are in order.

Traffic control came about for reasons of safety. The earliest control devices were hand signals, sometimes mistaken by early motorists as friendly greetings. The first traffic light was installed in Cleveland in 1914,[2] a fact ignored by the city of Detroit, which has enshrined its own "first traffic light" at the Ford Museum in Dearborn. Traffic lights proliferated, and their use evolved in two ways, schematically shown in Figure 1. The evolution was intended to accomplish yet another objective of traffic control, decreasing the inconvenience of unnecessary stops and delays sometimes caused by traffic lights.

The first notable innovation in traffic control was the synchronization of a string of lights to provide uninterrupted driving through them, preferably in both directions. The first such "progression" system was installed in Salt Lake City in 1918.[2] Then came the concept of the "multidial" system, one that could react to changing patterns of traffic in the course of the day. The first multidial systems were open-loop systems. They generally had three options of synchronization: one for morning rush periods, one for afternoon rush periods, and one for average conditions. Each of the three options was designed to accommodate expected traffic demands, and the transition from

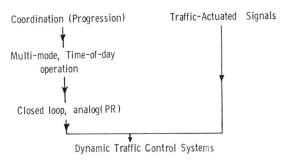

FIGURE 1. Evolution of traffic control systems.

one dial to another was carried out at fixed times during the day. The next step was to make the transition from one synchronization scheme to another responsive to actual measurements of traffic. The best known example of such a system was the PR System installed in Baltimore in the early 1950s. In that system, the transition from one synchronization scheme to another was made on the basis of measurements of traffic at some key points in an arterial network.

The right-hand side of Figure 1 shows another branch of the evolution tree. Rather than using a traffic light with a fixed cycle, insensitive to actual traffic demands, we started responding to signals produced by cars approaching the intersection. Some of these traffic-actuated signals are quite sophisticated, responding to variable traffic demands on one or both legs of the intersection. All of them control the intersection in isolation, although the traffic stream itself moving from one intersection to another may establish some semblance of progression in certain types of these signals.

Both of the preceding concepts, progression and traffic-actuation of individual intersections, have been tied together in modern dynamic traffic control systems run by digital computers. Computerized systems borrowed heavily from the principles of operation of past systems. Some of these principles and engineering practices have been scrutinized using techniques of system optimization and control theory. Other applications of control theory go beyond current engineering practices in an attempt to provide a theoretical foundation for the traffic control systems of the future.

In this chapter we discuss some of these applications of system optimization to traffic control. We begin with a discussion of the main objectives of traffic control, and principles of operation of traffic control systems, which is given in Section II. Section III contains a review of signal synchronization schemes. Section IV gives a discussion of some heuristic schemes for the control of critical intersections or systems of such intersections. Section V contains a review of work on oversaturated traffic systems, that is, on systems which become overloaded during a rush period. And Section VI gives a discussion of some general principles of control of critical traffic links, such as express-ways and tunnels.

II. OBJECTIVES AND PRINCIPLES OF OPERATION OF TRAFFIC CONTROL SYSTEMS

The general objectives of traffic control are to improve safety and to increase the comfort and convenience of drivers. This general statement of noble intention must first be reduced to quantifiable objectives. Ideally one would like to set up the traffic control problem as a system optimization problem, by defining an *objective function* and seeking an optimum (maximum or minimum) for this function. This function will generally depend on certain controllable parameters of the system; for example, the light cycle, the *split* (allocation of green time to different traffic streams), *offset* (time phase of cycles) of different lights, and so on.

Many criteria of performance have been suggested in traffic control, among which are the following:

1. Average delay per vehicle,
2. Maximum individual delay,
3. Percentage of cars that are stopped as they go through the system,
4. Average number of stops before a vehicle goes through the system (or through an intersection),
5. Throughput of the system (or individual intersections),
6. Travel time through the system (average of maximum).

Criterion 1 is the most widely used, and it is not always compatible with criterion 2. It is sensible to use criterion 1 as a primary objective function with 2 used by way of a constraint on the maximum individual delay. This is not only for humanitarian reasons, but also for a very practical one: Any design that penalizes a few drivers very much may induce changes in their driving habits, which in turn may destroy the design. Criteria 3 and 4 are dictated by the known aversion of drivers to stops. Stops are both a nuisance and an expense because of excessive gasoline consumption and wear and tear on cars. In addition, in the last few years we have become increasingly aware that stops tend to increase car-induced air pollution. Criterion 5 is particularly important during periods of peak traffic, when it is important to maximize the utilization of the traffic system. Finally, criterion 6 is very much related to criterion 1, and in fact it is frequently used in lieu of that criterion.

Sometimes an objective function is not explicitly related to the preceding criteria. An example is the through-band (or green-band) design of synchronization of traffic lights. In that design, one tries to align the green phases of a sequence of traffic lights with such time offsets as to allow uninterrupted flow through as many of them as possible to as many drivers as possible. The through-band design more or less maximizes the opportunity of driving through a string of traffic lights without stopping. Indirectly, it tends to reduce stops and delay—sometimes.

Among the traffic systems that have been investigated for the purpose of optimizing their operation are the following:

1. An isolated signalized intersection,
2. A system of a few intersections (two to five),
3. An urban street network controlled by traffic lights,
4. Critical traffic links such as tunnels, bridges, and sections of freeways.

For all these systems, one would like to have a methodology for optimizing their operation with respect to a given objective function, in response to observations concerning traffic inputs. A general theory of optimization of traffic systems, particularly with realistic stochastic traffic inputs, has been rather elusive. Exact optimization has been obtained only for a few simple systems with special traffic inputs. For most traffic systems we have relied heavily on heuristic control schemes coupled with suboptimization schemes for idealized versions of the real systems, or portions thereof. Let us briefly outline the general principles of operation which have been suggested for the four types of systems listed above and the types of optimization problems encountered in each case.

The single intersection operating below saturation levels is discussed in Chapter 2 of this book. In that chapter it is shown how one can select a fixed cycle that minimizes delay when the traffic input is a stationary stochastic process. In Section IV of this chapter, we discuss the case of an oversaturated intersection. What distinguishes the oversaturated from the undersaturated case is the time dependence of traffic inputs plus the fact that the instantaneous information concerning traffic inputs is not sufficient for optimization. Residual, long queues propagate the delay from one light cycle to another. Thus the optimization of the operation of the traffic light of an oversaturated intersection entails minimizing the delay to all the traffic which crosses the intersection during the rush period; that is, the period of oversaturation. A similar approach is discussed in Section IV for systems of oversaturated intersections, and a few other special oversaturated systems.

As was mentioned in Section I, the methodology of control of urban street networks has borrowed heavily from the traditional methodologies of progression and traffic actuation. Thus, a typical modern control system works on the following principle: One tries to synchronize as many traffic lights as possible in order to provide the opportunity of uninterrupted flow to as many cars as possible. The synchronization scheme is selected for average traffic conditions, and changes relatively infrequently, sometimes no more than three or four times during the day. In any case, the time scale of switching from one synchronization scheme to another is of the order of minutes or even hours. This type of control has been referred to as *macrocontrol* (or *major loop control*). The problem associated with macrocontrol is to find a good synchronization scheme for given traffic conditions.

Over and above macrocontrol, we frequently need to pay special attention to critical intersections. This is done usually at time increments of the order of one second, and has been referred to as *microcontrol* (or *minor loop control*). The problem associated with microcontrol is to find an algorithm for adjusting the traffic lights in response to variations of traffic demand, in order to minimize the delay in the neighborhood of the critical intersection. Both macrocontrol and microcontrol coexist in most modern, computerized traffic control systems. However, most systems in operation today depend largely on macrocontrol schemes. The degree of utilization of microcontrol schemes is limited by our lack of experience concerning the inner workings of traffic. Since microcontrol is expensive in instrumentation requirements, it is unwise to invest too much in it without reliable information about its efficacy.

The fourth class of traffic systems usually involves the need of maintaining free flow over critical traffic links, such as tunnels, bridges, and freeways, by regulating the traffic input into these critical links. Much of the methodology in this area has been developed empirically. Some models of input control have been cast in the framework of mathematical programming and control theory. These models, as well as the empirical solutions, are discussed in Section VI.

III. SYNCHRONIZATION SCHEMES

As was mentioned in Section II, the objective of a synchronization scheme for a system of traffic lights is to allow as many drivers as possible to go through the system with the minimum inconvenience. Generally, a measure of the inconvenience to the drivers is the delay they incur over and above the natural travel time through the system. Another measure, implicitly or explicitly taken into account, is the number of stops that a car makes as it goes through a system of traffic lights.

The oldest synchronization scheme, and still the one most widely used is the *maximum through-band* design shown in Figure 2. It consists of offsetting the

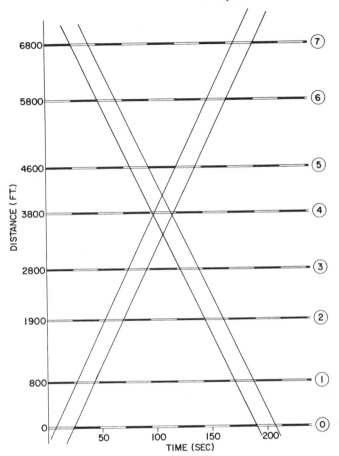

FIGURE 2. A typical time–space diagram showing the through-band design of progression.

beginnings of the green phases of successive lights with respect to each other in order to allow as many cars as possible to drive through these lights at some design speed without stopping. It is tacitly assumed that the path of these cars is not blocked by other cars, an assumption which is not always realized in practice. Traffic engineers have used nomograms, graphical techniques, and trial-and-error procedures for maximizing through-bands. In recent years, efficient computer algorithms for this purpose have been given by Morgan and Little[3] and by Brooks.[4] We shall discuss here the algorithm of Brooks, as an illustration of the general features of progression design.

It is assumed that the following parameters are given:

1. The split, or allocation of green time, for each intersection,
2. A range of acceptable speeds, $v_{min} \leq v \leq v_{max}$,
3. A range of acceptable cycles, $c_{min} \leq c \leq c_{max}$.

It is required to find a set of offsets, a cycle, and a speed such that the ratio of the sum of the through-bands in the two directions over the cycle, defined as the *efficiency* of progression, is maximized.

We first make the following observations:

1. If we have any progression design such as that of Figure 2, we can obtain an infinite number of alternate designs with the same efficiency by changing the scale of Figure 2. This means that the design speed v and the cycle c are scaled into λv and c/λ, where λ is the scaling factor.
2. Any progression design with unequal through-bands in the two directions can be obtained from one with equal bands by sliding the offsets of some key intersection in an appropriate direction. Thus, it is sufficient to design for equal bands in the two directions.
3. A progression design for unequal speeds in the two directions can be obtained by "shearing" the diagram of Figure 2 in the direction of the abscissa axis. Thus, the basic problem is reduced to designing for equal speeds in the two directions.

Using the above observations we can dispense with one of the variable ranges of the problems. For example, we can set the cycle at c_{min} and design for an expanded range of v; namely, $v_{min} \leq v \leq v_{max}(c_{max}/c_{min})$. It is further convenient to discuss the case of a fixed design speed, since covering the range of v is then a matter of a search over a single variable. It may be mentioned that certain observations, discussed by Brooks,[4] make this search an extremely simple one. We have now reduced the through-band design problem to the following:

PROBLEM. Given a design speed and a cycle, as well as the splits of all intersections, find the offsets that maximize equal through-bands in the two directions of an artery.

From symmetry, we find that the offsets must be such that the middle of the green phase of any intersection must occur simultaneously with the middle of either the red or the green phase of any other. In addition, it can be seen that if all travel times between intersections are integer multiples of the half-cycle, then one can have the maximum possible through-band in both directions, equal to the minimum green phase. There is an alignment of midphases such that a band of trajectories corresponding to the design speed drawn from any point in time during the minimum green phase must pass through the green phases of the other intersections [Figure 3(a)]. If the

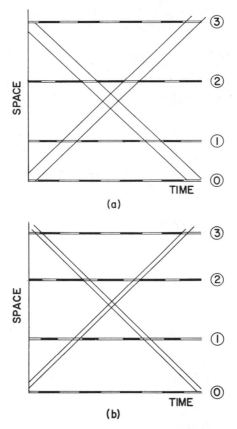

FIGURE 3. (a) The case of through-band equal to the minimum green phase. (b) The case of through-band smaller than the minimum green phase due to right or left interference at various intersections.

travel times between intersections are not multiples of the half-cycle, then the green-band emanating from the minimum green phase is generally blocked on one or the other side by some of the red phases of the other intersections [Figure 3(b)]. Depending on the alignment of midphases, any of these intersections produces either a left interference or a right interference; that is, they either block the left side or the right side of the green band emanating from the minimum green phase. The problem is then reduced to finding an alignment of midphases such that *the sum of the maximum left interference and the maximum right interference is minimized.* This is accomplished with the following finite algorithm:

1. Let d_i be the distances of the intersections from the intersection with the minimum green phase, taken as the reference intersection, g_i the green phases of these intersections, g_0 the green phase of the reference intersection, v the design speed, and c the cycle. Compute

$$t_i = \frac{d_i}{v},$$

$$l_i = t_i \bmod \left(\frac{c}{2}\right) + \frac{(g_i - g_0)}{2},$$

$$r_i = \left(\frac{c}{2}\right) - l_i + (g_i - g_0), \tag{1}$$

$$m_i = \mathrm{int} \frac{t_i}{(c/2)} \qquad i = 1, 2, \ldots, N,$$

where the symbols int and mod denote the integer part (largest integer contained) and modulo (residue after division of left-hand-side by right-hand-side number). The quantities l_i are the left interferences, and the quantities r_i the right interferences of the lights i on the band emanating from the reference intersection. Any given alignment of midphases corresponds to either the left or the right interference for each intersection.
2. Arrange the left interferences in descending order. Let L_i be this arrangement, and R_i the corresponding right interferences. One feasible solution is to select an alignment of midphases corresponding to all left interferences, in which case the bandwidth is

$$b_0 = g_0 - L_1. \tag{2}$$

3. We can now try and improve this solution by trading off the leading left interferences, up to and including the kth one, for right interferences. This is accomplished by shifting the offset of the corresponding intersections by $c/2$. The optimum bandwidth is obtained by maximizing over k the quantity

$$b_k = g_0 - L_{k+1} - \max_{j=1,k} R_j. \tag{3}$$

This maximization is obtained by a single pass, varying k from 1 to N. Furthermore, not all the values of k have to be considered, but only a subset corresponding to increasing values of R_j.

4. Once k is determined, the alignment of midphases is set by computing

$$a_i = (m_i + 1) \bmod 2 \qquad i \leq k,$$
$$a_i = m_i \bmod 2 \qquad k < i \leq N, \tag{4}$$

and setting the midphase of intersection i identical to that of the reference intersection when $a_i = 0$ and opposite when $a_i = 1$.

As was mentioned already, starting from the basic progression design just given, one can obtain others with equal efficiency but different speeds and/or through-bands in the two directions in order to accommodate different traffic demands along these directions. Little[5] introduced some additional flexibility by allowing small changes of speeds between different pairs of intersections, on the theory that drivers can learn to adapt themselves to such small changes. Allowing these changes could generally result in wider through-bands. At the same time, Little cast the progression design problem for an arterial network in the framework of mixed-integer linear programming. The objective function is a weighted sum of the bandwidths of all arteries in a network. Each bandwidth depends linearly on the offsets that are the continuous variables. The offsets must satisfy certain constraints involving certain integer variables. These constraints express the fact that adding relative offsets between successive pairs of intersections along a closed path through the network one must obtain an integer multiple of the cycle. Little's method handles in a general way any configuration of an arterial network. Its applicability is only limited by our inability to solve efficiently large mixed-integer linear programming problems.

We shall give here the formulation, following Little, of the problem already discussed; namely, the maximization of equal through-bands along a two-way artery. We assume that the cycle, splits, and travel times between intersections are given. By shearing of the time–space diagram between pairs of intersections, we can obtain one corresponding to equal travel times in both directions. Thus, the solution will be symmetric with respect to the two directions of traffic. Referring to Figure 4, we consider two intersections i and j, which may be separated by other intermediate intersections. We define the following variables (where all times are expressed in multiples of the cycle c):

$\rho_i =$ red phase of intersection i,

$t_{ij} =$ travel time from i to j $(= t_{ji})$,

$\phi_{ij} =$ time from center of the red phase at i to center of a red phase at j, where the two red phases are adjacent to a through-band and on the same side of the band $(= \phi_{ji})$,

$w_i(\bar{w}_i)$ = time from the right (left) side of the red phase of intersection i to the green band. (From symmetry, it is sufficient to consider a through-band in one direction only.)

From Figure 4 we obtain the following relationships:

$$\left(\frac{\rho_i}{2}\right) + \bar{w}_i + t_{ij} - \bar{w}_j - \left(\frac{\rho_j}{2}\right) = \tfrac{1}{2}m_{ij},$$

$$\left(\frac{\rho_j}{2}\right) + w_j + t_{ij} - w_i - \left(\frac{\rho_i}{2}\right) = \tfrac{1}{2}m_{ij},$$

$$(5)$$

where

$$m_{ij} = 2\phi_{ij} = \text{integer}. \tag{6}$$

In addition, from Figure 4 we obtain the constraints

$$w_i + b \le 1 - \rho_i,$$

$$\bar{w}_i + b \le 1 - \rho_i. \tag{7}$$

The problem can be now stated as follows:

Find b, w_i, w_i, and m_{ij} in order to maximize b subject to the constraints given by Eqs. 5–7 plus the positivity constraint

$$b, w_i, \bar{w}_i \ge 0. \tag{8}$$

The above problem is a *mixed-integer linear programming* problem. The constraint that the variables m_{ij} be integer is the essential complication.

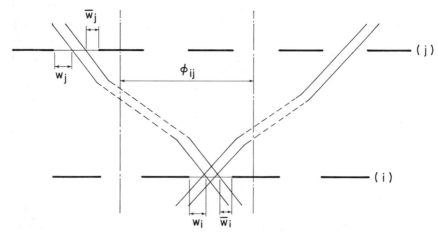

FIGURE 4. Definition of through-bands and associated parameters in Little's formulation.

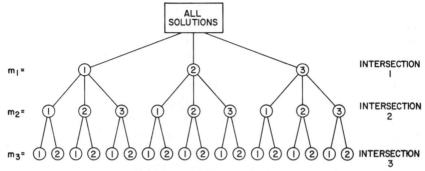

FIGURE 5. The "tree" of synchronization schemes used by Little in his branch and bound algorithm for finding the optimum scheme.

For any set of integer values for m_{ij}, the problem is an LP (linear programming) one that can be solved using standard methods and computer packages. For the mixed-integer linear programming problem, Little suggested a *branch and bound* technique for solving this problem. In effect, this technique provides a systematic search through a "tree" that can be formed by considering intersections one at a time and all the values of the variable m associated with each intersection and a reference intersection (Figure 5). Each node of this tree is associated with a value of the through-band, which can be obtained by solving an LP problem, assuming fixed values of m_{ij}, for a subset of the intersections. The branch-and-bound technique searches along the branches of the tree starting always from the node with the highest through-band, until all values of m_{ij} have been selected.

The method of Little offers few, if any, advantages for the simple problem discussed here. Its value lies in its generality and in the possibility of adding constraints in order to accommodate special requirements. Among the possible extensions of the above solution are the following:

1. The change of speeds between successive pairs of intersections may be allowed to range between two bounds. This is accomplished by introducing t_{ij} as additional variables that must satisfy appropriate constraints.
2. The through-bands in the two directions may be required to have any desired ratio.
3. The cycle may be allowed to vary between given bounds.
4. The method can be adapted for networks of arteries. One only need to introduce appropriate "loop constraints," which must be satisfied by variables such as the w_i along closed loops in the network. Little suggests as objective function in this case the weighted sum of the through-bands along all arteries in the network.

All through-band designs have one basic deficiency, namely, that they presume relatively empty streets. Everyone is familiar through personal experience with progression designs that get clogged because queues of cars block the path of cars through successive green lights. This situation has been discussed by this author,[6] and is illustrated in Figure 6. In the absence of queues, the ideal offset would be such that the green phase of intersection M starts d/v seconds after that of intersection N, where v is the speed of a moving platoon of cars. If, however, there is a queue of length l before intersection M, then the onset of green at M induces a starting wave that propagates toward the end of the queue with speed v_1, and reaches this end after time l/v_1. If the leader of the platoon stopped at intersection N is to reach the end of the queue in front when it is moving, he must start t_1 seconds before the green phase at M, where t_1 is given by

$$t_1 = \frac{d-l}{v} - \frac{l}{v_1}. \tag{9}$$

It is seen from Eq. 9 that t_1 becomes negative when l becomes sufficiently large, leading to a situation which has been referred to by some as "reverse progression." Whatever the apellation, the point is that adjustments must be made for the existence of stopped queues. This has been done, directly or indirectly, in some of the synchronization schemes suggested in recent years, such as the San Jose design by Chang,[7] and the TRANSYT method of Robertson.[8] These methods are effectively simulating the motion of cars

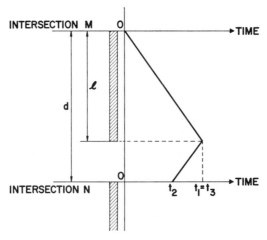

FIGURE 6. The effect of queues on progression design.

through a network, evaluating an objective function that measures the inconvenience to drivers, and then zeroing on a good set of offsets, through a search over the domain of allowable offsets. Differences between these schemes are in aspects of modeling of the movement of cars, such as platoon dispersion, and the method of search used.

The Chang method[7] was developed in the course of a joint study by the city of San Jose and IBM, which led to the development of the IBM 1800 traffic control system.[9] Chang defined an objective function *delay*, which could be computed as a function of the offsets of the traffic lights in a system by effectively simulating the movement of traffic through the system. Chang starts with an arbitrary (but reasonable) set of offsets and improves them by using a sequential search procedure. At each step of the procedure, the delay to all cars is computed and the set of offsets is changed in a way that reduces the overall delay. In simulating the flow of traffic through the system, Chang makes the following assumptions:

1. All events have a common period that is the common cycle of all intersections.
2. Cars move at constant speed until they meet a stopped queue. Then they contract to a higher density at constant rate, and in effect they do so by decelerating instantaneously to zero speed.
3. When the front of a queue can move again, cars accelerate instantaneously to a constant design speed associated with a constant density of a moving platoon.
4. Platoons of cars have a uniform density corresponding to a moving platoon and another higher density corresponding to a stopped queue.
5. There are no turning movements in the system. Actually, Chang made allowance for such turns by assuming that a fixed percentage of the traffic *sunk* at each intersection. These sinks of traffic, corresponding to turning movements, were roughly offset by *fixed* sources generating constant inputs of traffic at appropriate intersections.

From assumption 2, it follows that the rate of flow of cars at points between intersections are equal to flows at intersections, but with a time shift. Specifically the flow at point x between intersections i and j, Figure 7 can be obtained from the flow at i, provided that any queue at j does not reach up to x. This flow at x is given by

$$q_x(t) = q_i\left(t - \frac{d_{ij} - x}{v_{ij}}\right), \tag{10}$$

where $q_i(t)$ is the flow at i, d_{ij} is the distance from i to j, and v_{ij} the design speed from i to j.

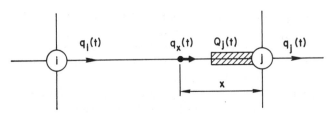

FIGURE 7. Queue formation according to Chang's model.

The number of cars queued at an intersection j, $Q_j(t)$, is now obtained using the model of Eq. 10. We assume that each car stopped in a queue at j occupies a length l_j. Then, if $Q_j(0)$ is the queue at j at time 0, the queue size at time t is given by

$$Q_j(t) = Q_j(0) + \int_0^t \left[q_i\left(\tau - \frac{d_{ij} - l_j Q_j(\tau)}{v_{ij}} \right) - q_j(\tau) \right] d\tau, \qquad (11)$$

where it has been tacitly assumed that the length of the queue, $l_j Q_j(\tau)$, remains smaller than d_{ij} in the time interval $(0, t)$. Another assumption implicit in Eq. 11 is that the dissipation of a queue at j takes place by the entire queue moving toward j to fill the space vacated by a departing vehicle. In effect, this implies that the starting wave propagates instantaneously to the end of the queue. This can be seen by differentiating Eq. 11 and setting $q_i(t) \equiv 0$. In that case, the length of the queue at j, $L_j(t)$ is found to satisfy the differential equation

$$\frac{dL_j(t)}{dt} = -l_j q_j(t). \qquad (12)$$

Chang points out that a more realistic model for the length of the queue might be given by an equation of the type

$$\frac{dL_j(t)}{dt} = l_j \left[q_j\left(t - \frac{d_{ij} - L_j(t)}{v_{ij}} \right) - q_j\left(t - \frac{L_j(t)}{v_{ij}} \right) \right]. \qquad (13)$$

However, he rightly points out that in view of the many simplifications of the model of traffic flow he has used, a refinement such as that of Eq. 13 is not warranted.

The model is completed by deriving expressions for the rate of flow at intersection j. Let I_j be an indicator function for Q_j, defined as

$$I_j(t) = \begin{array}{ccc} 0 & \text{if} & Q_j(t) = 0 \\ 1 & \text{if} & Q_j(t) > 0. \end{array} \qquad (14)$$

Then the flow at j is given by

$$q_j(t) = \begin{cases} 0 & \text{if the signal is red} \\ S_j I_j(t) + (1 - I_j(t)) q_i\left(t - \dfrac{d_{ij}}{v_{ij}}\right) & \text{otherwise,} \end{cases} \tag{15}$$

where S_j is the *saturation flow* at j, that is, the maximum rate of flow at the front of a stopped queue. The amber phase is neglected in Eq. 15. It may be considered as part of either the red or the green phase, presumably depending on driving habits.

The preceding definitions pertain to internal intersections in a network. At the boundary intersections, one must specify sources according to statistical information. From the set of equations given, one can then proceed to derive the sizes of all queues $Q_j(t)$ and using these sizes compute an objective function given by

$$D = \sum_j \int_0^c Q_j(t)\, dt, \tag{16}$$

where c is the cycle and $Q_j(t)$ are periodic solutions of time with period c. The quantity D is a measure of the time that cars spend in stopped queues over the entire system, that is, a measure of the delay to cars caused by the traffic lights. A reasonable objective is then to minimize D.

The *delay* D depends on the traffic light settings, and specifically on the offsets of these lights relative to a reference intersection. Chang proposed a heuristic search procedure over the admissible range of offsets. The procedure consists of a course search followed by a fine search. During the course search the system may be divided into subsystems in order to facilitate the search. The solution of the course search is used as a first approximation for the fine search. The reader familiar with nonlinear programming problems will understand the rationale of this procedure. The function D is a nonlinear, periodic function of the offsets. A fine search procedure may yield a local minimum that is far off from the global minimum. It is important, therefore, to obtain a fairly good starting point for the fine search. Chang proposed, and used, the through-band design for individual arteries as a course search. This amounts to finding the best of 2^{n-1} possible alignments of midphases of green or red of $n - 1$ intersections with respect to a reference intersection.

The TRANSYT method of Robertson[8] is conceptually similar to that of Chang in that it simulates the movement of traffic through a network and seeks a traffic light setting that minimizes a weighted sum of delay and stops. The TRANSYT method assumes a model of traffic flow that includes the effects of *platoon dispersion*, that is, the change in the spatial distribution of a group of vehicles due to the differences of their speeds. TRANSYT also allows for some deviations from fixed splits. In addition to optimizing with respect to the offsets, it tries to improve the traffic further by varying the allocation of green at each intersection.

The TRANSYT procedure comprises a model used to calculate delays and stops and a "hill-climbing" procedure for minimizing an objective function, which is given by

$$D = \sum_{i=1}^{N} (d_i + Kh_i), \tag{17}$$

where d_i is the average delay per hour on the ith link of a network containing N links, h_i the average number of stops per second on this link, and K a weighting factor.

The model of traffic movement consists of generating a pattern of flow on any network link by manipulating three types of flow patterns for each direction of flow

1. The IN pattern, which is the flow of traffic past the downstream end of a link if the traffic is not impeded by a traffic light,
2. The OUT pattern, which is the actual traffic flow past an intersection into the next link,
3. The GO pattern, which is the flow that is obtained if there is a residual queue upstream of the intersection during the entire green phase.

The OUT pattern of one intersection generates the IN pattern at the next downstream intersection, after allowing for dispersion (spreading out) of the platoons of cars. A typical IN pattern is shown in Figure 8, in the form of a histogram corresponding to flow during any one of 50 equal fractions of a light cycle. (The number 50 is used because most lights in Great Britain have a light cycle of 50 sec.) It may be seen that from the position of the green phase relative to the IN pattern one can compute both the delay and the number of stops at an intersection, and consequently the objective function D. A minimum for this function is then obtained by the following hill-climbing procedure. The offset and/or split of a single intersection is changed by a small number of units and the variation of D is computed. The change in offset and split is continued until a (local) minimum of D is obtained. The process is repeated varying the offset and split of the other intersections in a preselected sequence, completing one iteration. The procedure is stopped after several iterations, when the marginal improvement begins to be very small.

Another method of signal synchronization is the combination method given by Hillier and by Whiting.[10,11] The combination method, like many others, assumes a common cycle and given fixed splits at each intersection.

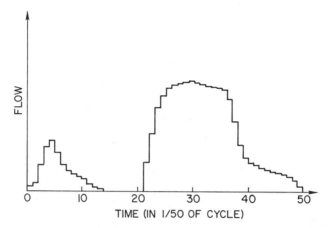

FIGURE 8. A typical IN flow diagram of the TRANSYT program.

In addition, it is based on two assumptions that are only approximately satisfied in real life:

1. That the amount of traffic at each intersection does not depend on the traffic light settings,
2. That the delay along a given link of a network depends only on the settings of the lights at the two ends of the link.

The combination method is based on the observation that, on the basis of the above assumptions, links can be combined in series or in parallel in computing and minimizing delay. Let us explain briefly this notion of combination.

Consider a pair of intersections A and B (Figure 9) connected by two traffic links. Assumption 2 above states that the delay of traffic along link 1 depends only on the offset between A and B. Let this delay be given by a function

$$D_1^{AB}(z_{AB}) \qquad 0 \leq z_{AB} \leq c, \tag{18}$$

where z_{AB} is the offset of B with respect to A. It is sufficient to define D_1^{AB}, only in the interval $(0, c)$, where c is the cycle, since it is a periodic function

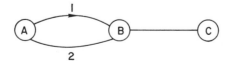

FIGURE 9. Combination of links in parallel and in series.

of z with period c. We define a similar function, $D_2^{AB}(z_{AB})$, for the delay along link 2. Since the two delays are independent of one another, they can be added in computing the total delay of the network, which includes A and B. We can then proceed by replacing the two links by a single link associated with a delay function:

$$D^{AB}(z_{AB}) = D_1^{AB}(z_{AB}) + D_2^{AB}(z_{AB}). \tag{19}$$

Consider next two links in series connecting three consecutive intersections A, B, and C. We define the measurable delay functions

$$
\begin{aligned}
D^{AB}(z_{AB}) \qquad 0 \le z_{AB} \le c, \\
D^{BC}(z_{AB}) \qquad 0 \le z_{BC} \le c.
\end{aligned} \tag{20}
$$

We now can compute the delay of traffic moving from A to C, D^{AC}, as a function of the offset of the light C with respect to the light A, z_{AC}. This, of course, depends also on the relative offset of the light B. We pick the offset at B in such a way as to obtain the minimum delay from A to C for any given value of z_{AC}. We do this by defining

$$D^{AC}(z_{AC}) = \min_{0 \le z_{AB} \le c} [D^{AB}(z_{AB}) + D^{BC}(\tilde{z}_{BC})], \tag{21}$$

where

$$
\begin{aligned}
\tilde{z}_{BC} &= z_{AC} - z_{AB} \qquad \text{if} \qquad z_{AC} \ge z_{AB} \\
&= c + z_{AC} - z_{AB} \qquad \text{if} \qquad z_{AC} < z_{AB}.
\end{aligned} \tag{22}
$$

We can now replace the links AB and AC by a single link AC in computing the total delay in the network.

This combination of links in series and in parallel can be used to reduce a network into one with fewer links. Hillier observed that a network in the form of a ladder (Figure 10) could be reduced to a single link, making the

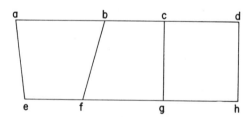

FIGURE 10. A "ladder" network which can be reduced to a single link by the combination method.

optimization procedure particularly simple, and also exact within the limitations of the design assumptions stated above. Allsop[12,13] has shown that any network can be "condensed" by the above process to an irreducible form in which at least three links meet at each node. As an example, the network in Figure 11(a) can be condensed into that of Figure 11(b) by combining links as follows:

AB and BC in series into AC,
AC and CD in series into $(AD)_1$,
$(AD)_1$ and AD in parallel into $(AD)_2$,
$(AD)_2$ and DE in series into AE,
EF and FG in series into EG,
GH and HI in series into GI,
IJ and JA in series into IA.

In combining links in series, Hillier suggests discretizing the interval $0 \leq z \leq c$ and searching for the minimum of Eq. 21 at discrete values of z. Typically, 50 discrete values are considered, since the most common cycle in Great Britain is 50 sec, and it is convenient to derive offsets expressed in seconds that can be easily implemented with available hardware.

The synchronization of lights for an irreducible network such as that of Figure 11(b) has been considered by Allsop,[12,13] who gave an iterative scheme for minimizing the delay in such a network by an appropriate selection of offsets. The scheme starts with a single link and proceeds to form larger and larger subnetworks in successive stages. At each stage, one estimates, for any combination of offsets at intersections on the boundary of the subnetwork, a

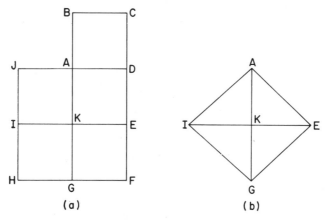

FIGURE 11. Reduction of a network (a) to its irreducible form and (b) by the combination method.

set of offsets of internal intersections that minimizes delay. The process proceeds until all links are included. It may be seen that the combinatorial nature of the process limits the size of the network that can be handled by a computer with any given memory size. As a rule of thumb, Allsop indicates that a memory size of 256K bytes (64K words) can handle up to 100 intersections.

We shall next discuss the SIGOP (signal optimization) program,[14] developed under the auspices of the U. S. Bureau of Public Roads by the Traffic Research Corporation. This program seeks to minimize, in a least-square sense, the deviation of the relative offsets of neighboring intersections from a set of ideal relative offsets that can be stipulated, computed, or measured. It is implicitly assumed that the delay is a periodic function of the relative offsets, as shown in Figure 12(a), with a minimum at some ideal value between $-c/2$ and $c/2$ (c = cycle), or this value incremented by a multiple of c. It is

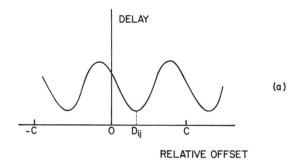

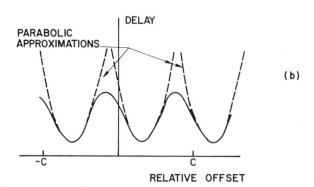

FIGURE 12. Contribution to the delay function used in SIGOP along a link ij between intersections i and j. (a) Exact contribution. (b) Parabolic approximation (Eq. 23).

further assumed that this ideal relative offset is independent of the offsets of the other intersections. This delay function may be approximated by parabolas with the same minima, as shown in Figure 12(b), leading to an objective function of the type

$$D = \sum_{ij} \alpha_{ij}(D_{ij} + \theta_i - \theta_j + M_{ij}c)^2, \tag{23}$$

where

$D_{ij} =$ *ideal* relative offset between intersections i and j,

$\theta_j =$ design offset of intersection j,

$\alpha_{ij} =$ link importance factor, weighing the traffic from i to j,

$M_{ij} =$ an integer value, usually ± 1 or 0, selected so that the relationship

$$\frac{-c}{2} \leq D_{ij} + \theta_i - \theta_j + M_{ij}c \leq \frac{c}{2} \tag{24}$$

is satisfied. The procedure for minimizing D is a combination of a Monte Carlo random starting-point routine and a direct search procedure. In effect, the multidimensional domain

$$0 \leq \theta_j \leq c \tag{25}$$

is divided into different regions corresponding to various choices of the M_{ij} satisfying Eq. 24. The SIGOP program searches for the global minimum of D within the domain (Eq. 25) by systematically obtaining local minima, through the following sequence of steps:

1. A random choice of θ_j is made satisfying Eq. 25, and the corresponding region is determined by finding all the M_{ij} satisfying Eq. 24.
2. The minimum of D corresponding to the region determined in step 1 is found by differentiating Eq. 23 and solving the resulting system of linear equations in θ_j.
3. Step 2 is repeated after changing each M_{ij} by ± 1 until a local minimum is obtained, which is such that a further change of any M_{ij} by ± 1 would yield a higher value of the minimum of step 2.

Steps 1–3 constitute one Monte Carlo "game." New games are started with step 1, and new values of local minima are obtained, until one is sufficiently certain that the global minimum is among all the local minima obtained. How one becomes sufficiently certain of having found the global optimum is a matter of judgment requiring a great deal of art in the use of SIGOP. It should be pointed out that an exhaustive search of all the regions of the domain (Eq. 25), corresponding to all values of M_{ij}, is prohibitively expensive for any network of moderate size. For example, it has been estimated[15] that a square grid with 25 signalized intersections with all streets two-way corresponds to approximately 10^{28} regions of the domain (Eq. 25).

We have mentioned some of the principal contributions to the problem of synchronizing traffic lights, with an emphasis on those which have been actually used. The subject of light synchronization has been one of the liveliest ones over the past eight years or so. In addition to the contributions already discussed, there have been many others, among them the following.

Buckley et al.[16] have given methods for maximizing through-bands for regular rectangular grids of one-way arteries. Gartner[17] has used arguments of graph theory to prove theorems concerning the optimum through-band design for a network. The theorems concern relationships between design parameters such as offsets, and are likened by Gartner to Kirchhoff's laws for electrical networks. Bavarez and Newell[18] have obtained a rigorous optimization of through-bands for one-way arteries, for a variety of delay-oriented objective functions involving both delay to drivers and number of stops. In effect, Bavarez and Newell consider the end effect of starting up a platoon at the beginning of the artery. According to their analysis, for any time dependence of the traffic input at the entrance of the artery, there exists in general a synchronization scheme better than the one corresponding to a continuous through-band. It is possible, in general, to reduce the objective function by "juggling" the offsets of the intersections near the entrance of the artery destroying the continuous through-band design. The usefulness of this solution is limited by the fact that one is rarely given the exact time dependence of the traffic input into an artery.

Two recent papers by Okutani[19] and by Gartner[20] use dynamic programming[21] for optimizing the synchronization of traffic lights with respect to delay. Okutani assumes that the delay between pairs of intersections depends only on the relative offsets of these intersections. He then uses dynamic programming to obtain the optimum synchronization, by dividing the total network into sections which are treated as stages in a multistage decision process. He applies Bellman's principle of optimality to obtain a recursion relationship between the optimum design for n sections and for $n + 1$ sections. Okutani also suggests an alternative procedure utilizing the discrete maximum principle.[22]

Gartner[20] also considers portions of a network as stages in a multistage decision process. Each stage is small enough that an exhaustive search on a small number of independent variables is possible in optimizing the stage. Optimization consists of minimizing a suitable objective function associated with the delay and/or number of stops of the traffic. The multistage process is optimized using dynamic programming. One distinctive feature of Gartner's approach is that he suggests a procedure, based on graph-theoretical concepts, for optimizing the minimization process in terms of the required computational effort.

It is appropriate to ask at this point, what are the advantages of one synchronization scheme over another or even over a random pattern of traffic lights? We can give only a partial answer to this question because not all of the synchronization schemes suggested have been tested. Testing of some synchronization schemes was carried out in San Jose[9] and in Glasgow.[23] These tests indicate that some of the schemes that take into account, explicitly or implicitly, the formation of queues at the intersections are better than schemes of the through-band type with respect to minimizing delay. The possible reduction of delay by a good coordination scheme, compared to the traditional through-band design, may be as high as 15% for moderately heavy traffic. Moreover, a design that takes into account the existence of queues tends to avoid releasing the traffic at an intersection when the up-stream section of an artery is blocked by a stopped queue. Thus, one avoids the familiar blocking of an intersection by traffic that cannot move forward, but is caught in the middle of the intersection when the light changes. Thus, designs that take into account the existence of queues tend to improve the utilization of intersections, a fact particularly important during periods of moderate to heavy traffic.

IV. HEURISTIC CONTROL ALGORITHMS FOR MICROCONTROL

Single Intersections

The evolution of traffic-actuated traffic lights has depended heavily on empiricism. Typically, these lights have been designed to extend a green phase when demand is continuing in any given direction, up to a maximum green phase. The extension is often implemented in multiples of a unit extension. A fully actuated traffic light, that is, one that detects and responds to traffic in both competing directions, generally tends to give the maximum allowable green phase to both directions. This is in line with the desire to increase the utilization of the green phase by decreasing the time lost for clearing the intersection when the light changes.

Computer control of city streets has borrowed heavily from the experience with traffic-actuated signals in the implementation of microcontrol schemes for critical intersections. However, until the middle 1960s, no detailed analysis of the stability of operation of these lights, or attempt at optimizing them, had been made. One of the first contributions in this area was made by Dunne and Potts,[24] who proposed a heuristic control algorithm for a traffic-actuated intersection and gave an interesting method for representing and analyzing the operation of the traffic light. The representation consists of describing

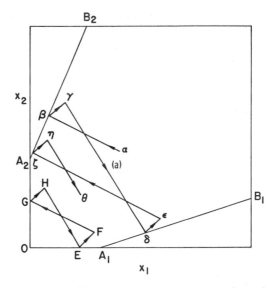

FIGURE 13. The Dunne–Potts representation of
queue behavior at an intersection.

the condition of the intersection in the state space (x_1, x_2), where x_1 and x_2 are
the sizes of queues of waiting vehicles along two competing directions of
traffic (Figure 13). If the variables x_i are allowed to be continuous, then the
condition of the intersection is represented in state space by a continuous
line. If, furthermore, the traffic inputs are assumed constant and continuous
and the flow rates are equal to (constant) saturation flows, then the condition
of the intersection is represented by a point moving along one of three
possible straight lines:

1. When the light favors direction 1, the state of the intersection (x_1, x_2)
   moves along the direction of the vector $(q_1 - s_1, q_2)$, where q_i are the
   input rates and s_i the saturation flow rates along the two directions.
2. When the light favors direction 2, the state of the intersection moves along
   the direction of the vector $(q_1, q_2 - s_2)$.
3. Finally, during the period of clearance between the above two cases, the
   state of the intersection moves along the direction of the vector (q_1, q_2).

In cases 1 and 2, it is assumed that there is a queue of cars waiting at the
intersection in both directions. When the queue in the favored direction is
exhausted, the corresponding axis $x_i = 0$ acts as a *barrier*, and the state of
the intersection moves along this barrier until the light is changed.

Dunne and Potts proposed and tested an algorithm which treats the lines
OA_1B_1 and OA_2B_2 (Figure 13) as reflecting barriers; that is, the light changes

as soon as the state of the intersection reaches either one of these two lines. Incidentally, when the reflecting barriers are the axes $x_i = 0$, then the Dunne–Potts algorithm is identical to what is known as the *saturation flow algorithm:* that is, the light changes only when the favored queue is just exhausted.

The Dunne–Potts algorithm yields a trajectory in state space such as the line $\alpha\beta\gamma\delta\epsilon\zeta\eta\theta$ (Figure 13). Under appropriate conditions it may yield a limit cycle, such as that shown by the line *EFGH* in Figure 13. The main results of Dunne and Potts are the following:

The algorithm is *stable,* that is, it results in bounded values of $x_i(\infty)$, when the intersection is undersaturated; that is, when

$$\frac{q_1}{s_1} + \frac{q_2}{s_2} < 1. \tag{26}$$

In that case, a limit cycle exists, with phases $g_{1\infty}$ and $g_{2\infty}$ given by

$$g_{i\infty} = \frac{2ly_i}{1 - Y} \qquad i = 1, 2, \tag{27}$$

where

$$y_i = \frac{q_i}{s_i} \qquad i = 1, 2,$$

$$Y = y_1 + y_2, \tag{28}$$

and l is the time lost during a phase change-over. The duration of the limit cycle is given by

$$c = g_{1\infty} + g_{2\infty} + 2l = \frac{2l}{1 - Y}. \tag{29}$$

It might be noted that the above results concerning the limit cycle are independent of whether or not the portions A_iB_i of the reflecting barriers coincide with the axes x_i, for $i = 1, 2$. However, as Dunne and Potts point out, the delay to the users does depend on the position of A_iB_i, being minimum when A_iB_i coincide with the coordinate axes, that is, in the case of the saturation flow algorithm.

Grafton and Newell[25] considered the transient behavior of a traffic-actuated traffic light and posed the problem of optimizing the operation of the traffic light by minimizing a delay function given by

$$D_\mu(x_{10}, x_{20}; P) = \int_0^\infty [x_1(t; P) + x_2(t; P)]e^{-\alpha t}\, dt. \tag{30}$$

In Eq. 30, x_1 and x_2 are the queue sizes in directions 1 and 2, x_{10} and x_{20} the initial queue sizes at time $t = 0$, P denotes a control policy, and μ is an index with the value 1 or 2 depending on whether the green light favors stream 1 or stream 2 at time $t = 0$. The exponential $e^{-\alpha t}$ is a discount factor that is introduced mainly in order to make the integral finite. (In practice, α is taken as very small, and the integral of Eq. 30 is evaluated as a power series in α, keeping only terms of order α^{-1} and α^0 and neglecting all positive powers of α.) The traffic inputs q_1 and q_2 are assumed constant.

The minimization of D_μ is accomplished as follows. First, the delay D is computed assuming an immediate application of the saturation flow algorithm, which is used as the *basis policy* P_0.

There are two possible types of variations that can be made in the basis policy—switch before a queue is empty or switch some time after a queue is empty. Any policy can be considered as the basis policy modified by such variations applied at some appropriate times. Thus, we are led to consider two *comparison policies*, the following:

Policy P_μ': At time $t = 0$, switch from the state (μ, x_{10}, x_{20}) to the state (μ', x_{10}, x_{20}), with $\mu' \neq \mu$ and pursue the basis policy thereafter.

Policy P_μ'': Allow the queue which is discharging from the state (μ, x_{10}, x_{20}) to become empty and stay empty for a time θ_μ, then switch and pursue the basis policy.

We can compute a delay difference between the policies P_μ', P_μ'' and the policy P_0. The comparison leads to the conclusion that there are domains in the state space x_1, x_2 where the basis policy is optimal. There are other domains, depending on the value of μ, for which one of the two comparison policies, P_μ' or P_μ'', is optimal.

The optimization is now completed by application of Bellman's *principle of optimality*,[21] which in this case is given by the equation

$$D_\mu(x_{10}, x_{20}; P_m)$$

$$= \min_P \left\{ \int_0^\tau [x_1(t; P) + x_2(t; P)]e^{-\alpha t}\, dt + e^{-\alpha \tau} D_{\mu'}(x_{10}', x_{20}'; P_m') \right\}, \quad (31)$$

where P_m and P_m' are optimal policies from the states (μ, x_{10}, x_{20}) and (μ', x_{10}', x_{20}'), respectively, and (μ', x_{10}', x_{20}') is the state which results at time τ from the policy P. Equation 31 expresses the recursive property of the optimal policy, namely, that whatever the initial state of the system and the initial decisions until time τ, the remaining decisions after time τ must constitute an optimal policy with respect to the state resulting from the first decisions.

Application of the principle of optimality, in conjunction with the results of comparison of the basis and comparison policies, allows one to determine the optimal policy for any initial conditions. Grafton and Newell[25] have given a complete description of the properties of the optimal policy which always tends to a saturation flow policy. Depending on the initial conditions, the transient character of the optimal policy may include a finite number of modifications of the P'_μ or P''_μ type described above which set up the system for the most economical transition into the steady-state, saturation flow policy.

A general approach for a good suboptimal control of a complex intersection serving time-varying traffic demands has been given by van Zijverden and Kwakernaak.[26] The approach is inspired by principles of optimization of stochastic control system given by Kwakernaak,[27] and consists of the following steps:

1. The instantaneous traffic situation at the intersection is observed and transferred to a model of the intersection that exists in an on-line computer.
2. The computer uses this information to estimate the effects of every possible sequence of light settings in the near future and selects the sequence that is best in some predefined sense.
3. The first setting of the sequence is transmitted to the local traffic, controlled, and implemented.
4. At the end of this first setting, steps 1 through 4 are repeated.

The implementation of the above control requires accurate information concerning the position of individual vehicles in the neighborhood of the intersection, which in principle can be obtained from two detectors per traffic lane. The control method has the advantages that it can be applied to arbitrarily complex intersection and is adaptive to changes in the traffic conditions. It gives good results so long as the predictive ability of the model of traffic movement is good. For this reason van Zijverden and Kwakernaak emphasize that the period over which the behavior of the intersection is predicted must be chosen to be precisely the period over which accurate prediction of vehicle behavior is possible.

There has been very little effort to test the preceding algorithms in practice, in part because of the prohibitive instrumentation requirements and in part because of a low expectation of overall benefit. Traffic engineers have preferred to rely on time-tested traffic-actuation schemes that, with minor modifications, have been implemented as microcontrol algorithms in computerized systems. The saturation flow algorithm was singled out for testing in Glasgow,[23] but apparently has not received wide acceptance there or anywhere else. The reason is that delay is not the whole story, particularly during undersaturated conditions. The saturation flow algorithm by definition guarantees that every car has to stop. However, it is well known that both individual drivers and traffic managers would accept a little extra aggregate delay if it prevented a few stops. Therefore, it seems reasonable that for undersaturated conditions the general practice will favor average light cycles longer than the minimum ones produced by the saturation flow algorithm.

Systems of Intersections

One of the first schemes for a fully adaptive control of a large number of intersections was given by Miller.[28] His algorithm is based on making a binary decision at time intervals of duration h, for every traffic light, whether to leave the light in its present state or change it. The decision is made by computing the delay reduction that would be accomplished if the traffic light were left unchanged.

Let us assume, for example, that at a N–S, E–W intersection the traffic light is now green along the N–S direction. If we keep the green phase for an extra h seconds, the reduction of delay for N–S traffic is estimated to be

$$\Delta D_{NS} = (a + r_{NS} + l_{NS})\left(\delta_N + \delta_S - q_N \frac{1 - (\delta_N/s_N)}{1 - (q_N/s_N)} - q_S \frac{1 - (\delta_S/s_S)}{1 - (q_S/s_S)}\right),$$

(32)

where a is the duration of the amber phase, r_{NS} an estimate of the N–S red phase, l_{NS} a lost time for start-up of the green phase, δ_N, δ_S the number of vehicles expected to cross the intersection during h seconds from the N and S approaches, respectively, q_N, q_S are the arrival rates of vehicles per h seconds along the two approaches, as s_N, s_S are the corresponding saturation flow rates.

By deferring the start of the E–W green by h seconds we impose some additional delay to the vehicles already there and those due to arrive before the queues are exhausted during the next green phase. This additional delay is estimated to be

$$\Delta D_{EW} = h\left(n_W + n_E + \sum_{i=1}^{k_W} q_W + \sum_{i=1}^{k_E} q_E\right),$$

(33)

where n_W, n_E are the vehicles already waiting at the W and E approaches, q_W, q_E the corresponding arrival rates, and k_W, k_E are the smallest integers such that

$$n_W + \sum_{i=1}^{k_W} q_W - \sum_{i=2+l_W/h}^{k_W} s_W \leq 0,$$

$$n_E + \sum_{i=1}^{k_E} q_E - \sum_{i=2+l_E/h}^{k_E} s_E \leq 0.$$

(34)

Equations 34 determine the number of intervals of h seconds during which there is still a residual queue after the onset of the next green phase. In these equations, s_W, s_E are the saturation flows along the W and E directions per h seconds, and l_W, l_E the lost times along these directions.

Combining Eqs. 32 and 33, we find that the net reduction of delay to all the vehicles around the intersection, if we extend the N–S green by h seconds, is given by

$$\Delta D = \Delta D_{NS} - \Delta D_{EW}. \qquad (35)$$

When $\Delta D > 0$, the signal should be left unchanged for h seconds, after which time the process is repeated.

A somewhat similar approach was used by Ross et al.[29] for a small system of intersections around a critical intersection. Both Miller and Ross et al. tested their schemes by simulation and found them beneficial, particularly in accommodating large fluctuations of traffic such as those caused by temporary interruption of flow. Both schemes would probably be improved by application of the familiar techniques of adaptive control suggested by van Zijverden and Kwakernaak[26] for a single intersection, namely, predicting for the next five periods and applying the best strategy for one period only. Neither one of the two methods has been tested as yet in a real system, partly because of the relatively extensive instrumentation and computation requirements and partly because traffic engineers tend to avoid large departures from a progression system.

A scheme involving small adjustments about a progression design that do not extend from one cycle to the next was one of four schemes tested at Glasgow.[23] Labeled EQUISAT, it consisted of equalizing the saturation levels along the two competing directions of every intersection, that is, making

$$\frac{q_1}{s_1 g_2} = \frac{q_2}{s_2 g_2}, \qquad (36)$$

where q_i, s_i, and g_i ($i = 1, 2$), are the traffic input rates, saturation flows, and green phases, respectively.

Another scheme is the one proposed by Koshi[30] for the city of Tokyo. This scheme is the most adaptive one proposed for a large traffic system. In effect, it is a departure from the practice of separating macrocontrol and microcontrol in that it introduces microcontrol over the entire system. However, it does it in such a way as to produce a continuously changing grand strategy for the system of lights. Like all other microcontrol schemes, the Koshi scheme involves an extrapolation in time over a period of the order of one light cycle. For each intersection, a test is made to determine whether or not a small advance or delay of the next phase would decrease the overall delay to the users. If so, this change in the offset is implemented. Thus, the Koshi scheme is a continuous application of a policy improvement scheme, taking account of the changing traffic conditions. One advantage of this scheme is that it circumvents the problem of transition from one grand strategy of synchronization to another. The transition is usually done in some *ad hoc* fashion requiring several cycles and has been known to cause disruptions that have discouraged traffic engineers from introducing too many synchronization options into a system. The continuous adaptation suggested by Koshi avoids the problems of transition, at the expense of higher instrumentation and computation requirements than those of the conventional microcontrol–macrocontrol approach.

All of the preceding schemes involve only the operation of standard lights, which can only allocate a green phase to a particular traffic stream but do not give directional instructions. It has been the long-standing belief of this author that a successful attack on the traffic problem must ultimately involve a dynamic allocation of traffic facilities through *route control*. The possible benefits from route control are illustrated by an example.

Consider a network of one-way streets (Figure 14) carrying N–S and E–W traffic, with all intersections operating near saturation levels. A turning movement q along the path ABC will be in conflict with N–S traffic. In order to prevent queueing, the green phases of the N–S streets along BC must be reduced by an amount $\epsilon = q/s$, where s is the *saturation flow* along BC (i.e., the maximum flow rate per unit of green phase time). As a result, all intersections to the right will be underutilized, and the total throughput of the system will be reduced by

$$(\Delta Q) = \epsilon \sum_{j=1}^{n} s_j, \qquad (37)$$

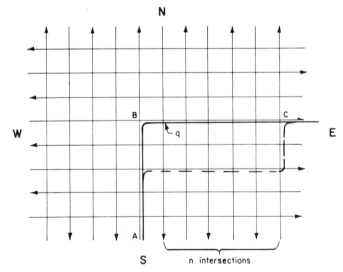

FIGURE 14. Possible benefits from route control.

where s_j is the saturation flow along the n N–S streets. It may be seen, however, that by splitting the streams q into two equal parts, one of them proceeding along the path shown by dashed lines, we can reduce the conflict by one-half along $n - 1$ of the N–S streets. This results in regaining half of the lost throughput along these $n - 1$ streets, or about half of the ΔQ shown in Eq. 37.

The subject of route control is discussed in some detail in Section V, where a model is proposed for congested traffic networks. Here, we may point out that congestion may sometimes be prevented by a judicious use of limited route control in the neighborhood of a critical intersection. Some examples of such route control have been given by Gazis and Potts.[31] The basic principle involved is that traffic can be fanned out and made to utilize some partially unused streets in the neighborhood of the point of congestion. This principle, together with appropriate synchronization of the traffic lights, may be used to produce green through-bands properly interwoven in a three-dimensional time–space domain (with two dimensions of space and one of time), so that large volumes of traffic may be passed through the system without stoppage. The simplest intersection complex of this type is the diamond complex shown in Figure 15. Theoretically, it can handle a continuous stream of traffic along the directions AB and BA, and platooned streams of traffic along CD and DC, with the platoons separated by gaps equal to their length. An extension of this system is shown in Figure 16 and can handle continuous streams of traffic in all four directions, AB, BA, CD, and DC. Another simple route control configuration given in Ref. 31 is shown in Figure 17. Theoretically, it can handle platoons in all four directions, AB, BA, CD, and DC, which occupy 75% of the roadway upstream of the intersection complex. Two-thirds of the traffic in each direction are guided through the middle intersection, which has a 50% split. The remainder is routed around the peripheral one-way loop. A proper coordination of the

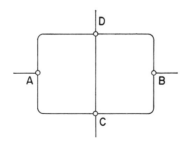

FIGURE 15. The basic principle of route control near points of congestion.

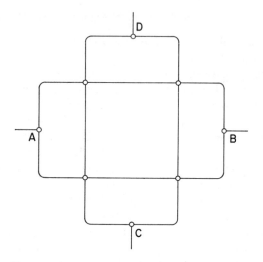

FIGURE 16. A complex of intersections which, with proper route control, can permit un-interrupted flow along both *AB* and *CD*.

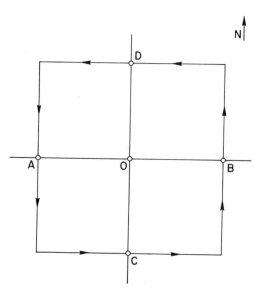

FIGURE 17. Utilization of a peripheral loop around a critical intersection, which can increase the flow along *AB* and *CD* by 50 %.

lights allows every user to move without delay, and without weaving, and provides a theoretical increase of throughput, compared to the single intersection 1, by 50%. The preceding discussion allocates this gain on a 50–50 basis between the N–S and E–W directions. However, different percentages of allocation can also be implemented, possibly adaptively, in response to traffic measurements. Also, the above scheme can be extended to a complex involving a peripheral loop around a group of intersections, for example, two. The potential benefit from the utilization of the peripheral loop decreases as the number of internal intersections increases.

V. OVERSATURATED SYSTEMS

In the preceding two sections, we have considered traffic systems in which traffic is not so heavy that it causes excessive queueing at any one point. We shall now consider oversaturated systems, that is, systems which become overloaded during a period of heavy demand, which will be referred to as the "rush period." We begin with the simplest case of a single intersection.

Single Intersection

Consider an isolated intersection serving n competing traffic demands. Cars arrive at rates $q_j(t)$ $(j = 1, \ldots, n)$, and are discharged during the appropriate green phases at a maximum rate, the saturation flow s_j, as long as there is a queue. When the queue is exhausted, the discharge, or service, rate is equal to the arrival rate. When the demand is low, the queues that are developed during the red phases are served completely during the green phases. The average green time required for serving all the cars arriving during a cycle c is

$$g_j = \frac{\bar{q}_j c}{s_j},$$

where $\bar{q}_j$ is the average arrival rate during the cycle. The standard rule-of-thumb used by traffic engineers has been to divide the total available green time in proportion to the g_j, allowing the same degree of saturation along all competing directions. Accordingly, the effective green phases g_j are selected so that they satisfy the relationships

$$\frac{\bar{q}_1}{s_1 g_1} = \frac{\bar{q}_2}{s_2 g_2} = \cdots = \frac{\bar{q}_n}{s_n g_n}. \tag{38}$$

The case of an undersaturated intersection has been discussed in Chapter 2, which contains a discussion of Webster[32] and others on the problem of optimizing the operation of a traffic light at an isolated intersection. As mentioned by Webster, the rule of equalizing the degree of saturation

(Eq. 38) is a fairly good one since it approximately minimizes the average delay of the users of an undersaturated intersection.

Let us now assume that the demand at the intersection increases, so that a stopped queue cannot be completely dissipated during a green cycle. For simplicity, we shall assume in the sequel that there are only two competing traffic streams at the intersection.

When q_i increase so that

$$\frac{q_1}{s_1} + \frac{q_2}{s_2} > 1 - \frac{L}{c}, \tag{39}$$

where L is the total lost time for acceleration and clearing, we have the case of oversaturation.[33] The rule of Eq. 38 may still be used, but it is meaningless and does not necessarily minimize the delay to the users. In this case, queues build up during a rush period and are completely served only at the end of the rush period. The point of view taken here is that one must try to minimize not the delay per cycle but the aggregate delay of both streams during the entire rush period. The situation is illustrated by Figure 18, in which the cumulative arrival curve Q and cumulative service curve G are plotted versus time. The quantities Q and G are defined by

$$Q(t) = \int_0^t q(\tau)\,d\tau, \qquad G(t) = \int_0^t \gamma(\tau)\,d\tau, \tag{40}$$

where $\gamma(\tau)$ is the average service rate, and the time origin is taken at the onset of the rush period. The service curve is actually such as the sawtooth curve of Figure 18 due to the succession of red and green phases, but will be

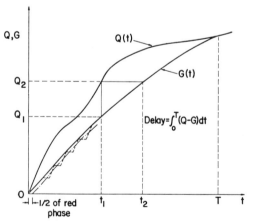

FIGURE 18. The oversaturated intersection—definition of the basic variables.

drawn as a smooth curve for simplicity. The area between the curves Q and G is a measure of the total delay to the users of the intersection during the rush period; their vertical distance is the effective size of the queue, and their horizontal distance, such as $(\tau_2 - \tau_1)$, a measure of the delay to an individual who demands service at time τ_1. The service rates are not entirely arbitrary, but must satisfy some constraints. This is so because the length of the green phase must generally lie within certain limits. Too short green phases are impractical or useless, and too long green phases are unacceptable to the stopped drivers of the cross traffic, who often assume that the light has failed and must, therefore, be ignored. If g_{min} and g_{max} are the bounds of the green phases, then the service rates γ_j satisfy the constraints

$$g_{min} \leq g_j = \frac{c\gamma_j}{s_j} \leq g_{max}. \tag{41}$$

The problem of optimizing the performance of the traffic light will now be formulated as an optimal control problem. The service rates are determined by the split of the available effective duration of green. Because of Eq. 41, the split may vary between an upper and a lower bound, defining a *control region*. Any acceptable service strategy involves time-dependent $\gamma_j(t)$ satisfying Eq. 41, and also the equation

$$\frac{\gamma_1(t)}{s_1} + \frac{\gamma_2(t)}{s_2} \leq 1 - \frac{L}{c} \quad (= \text{constant}), \tag{42}$$

where the sign of equality applies when sufficiently long queues exist in both directions to supply saturation flow during the entire length of the effective green phases. If, now, the minimization of aggregate delay is selected as the criterion of efficiency of operation of the traffic light, then the control problem may be stated as follows:

PROBLEM. Minimize the delay function

$$D = \sum_{j=1}^{2} \int_0^T [Q_j(t) - G_j(t)] \, dt, \tag{43}$$

where $\gamma_j(t)$ are subject to Eqs. 41 and 42, and the time limit T is defined by the equation

$$G_j(T) = Q_j(T). \tag{44}$$

It is tacitly assumed that after the time T, the capacity exceeds the demand.

The solution of the stated problem, following Ref. 34, is the following (see Figure 19):

There exists a time T that is the earliest possible end of oversaturation. This time is determined by assuming a single setting for the traffic light for

$0 \leq t \leq T$, that is, constant γ_j, such that both queues are dissolved simultaneously at time T. Thus the (constant) $\hat{\gamma}_j$ and T are determined from the system of three equations

$$\frac{\hat{\gamma}_1}{s_1} + \frac{\hat{\gamma}_2}{s_2} = 1 - \frac{L}{c},$$

$$\hat{\gamma}_j T = Q_j(T) \qquad j = 1, 2, \tag{45}$$

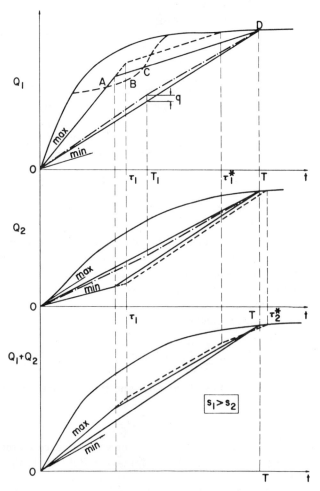

FIGURE 19. A graphical derivation of the optimum control of an oversaturated intersection.

where $Q_j(t)$ are known functions of t. The single-setting strategy is not the only one that dissolves both queues at the same time T, nor does it constitute the control that minimizes the aggregate delay. We can construct multistage service curves subtended by the straight lines corresponding to the single setting, all of which dissolve both queues at time T. The dash–dot lines of Figure 19 show such a strategy. Up to time T_1 we give stream 1 a little more green than that corresponding to $\hat{\gamma}_1$. We take away some green from T_1 until T, so that queue 1 is exhausted at time T. It can be proved easily that queue 2 will also be exhausted at time T. At time T_1, queue 1 is decreased, with respect to the single-setting strategy, by the amount q, say. Queue 2 is increased by the amount $(s_2/s_1)q$. Assuming, without loss of generality, that $s_1 > s_2$, we find that we have thus reduced the total delay by the amount

$$\delta D = \tfrac{1}{2}\left(1 - \frac{s_2}{s_1}\right)qT. \tag{46}$$

We can thus keep trading off delay until we construct the optimum strategy, which is a two-stage operation, in general. During the first stage, stream 1 is served with the maximum green phase and stream 2 with the minimum. During the second stage, the service rates are reversed, the minimum green phase given to direction 1 and the maximum to direction 2. The switch-over point is given by

$$\tau = \left(\frac{c}{s_1}Q_1(T) - g_{\min}T\right)(g_{\max} - g_{\min})^{-1}, \tag{47}$$

with T and $Q_1(T)$ satisfying Eqs. 45. The switching from one extreme of the controlable variable (green phase) to another is generally referred to as *bang–bang control* in control theory, a term somewhat less than reassuring when used in connection with traffic.

From the preceding discussion, it can be seen that the exact shape of the demand curves, near the beginning of the rush period, is not important, as long as the service curves do not intersect them. If the demand curves can be approximated reasonably well by straight lines near the end of the rush period, we can obtain simple expressions for the end of the rush period and the switch-over time. Let

$$Q_j(t) = A_j + B_j t. \tag{48}$$

Then

$$T = \frac{A_1 s_2 + A_2 s_1}{-(B_1 s_2 + B_2 s_1) + s_1 s_2 (1 - L/c)},$$

$$\tau = \frac{(A_1 b - A_2 a) + u_{\min}(\lambda A_1 + A_2)}{(\lambda a + b)(u_{\max} - u_{\min})}, \tag{49}$$

where

$$a = B_1,$$

$$b = B_2 - s_2\left(1 - \frac{L}{c}\right),$$

$$\lambda = \frac{s_2}{s_1}, \tag{50}$$

$$(u_{min}; u_{max}) = \frac{s_1}{c}(g_{min}; g_{max}).$$

References 33 and 34 contain discussions of other points concerning the single oversaturated intersection, such as the following:

1. The optimum light cycle, determined by balancing the additional delay due to the pulsating size of the queues against the reduction of delay due to increased efficiency associated with long cycles.
2. Exceptions from the preceding bang–bang type of control, due to special character of the demands and the resulting queues.
3. A formulation and solution of the problem using Pontryagin's maximum principle.[35] While Pontryagin's method is not always the ideal one for handling problems of control theory, it is quite instructive in the present case, and for this reason it will be given here.

Let us refer to Figure 20, which shows again the cumulative demand and service curves versus time for the two competing demands. The queue sizes x_1 and x_2 are the *state variables* satisfying the differential equations

$$\frac{dx_1}{dt} = q_1(t) - u,$$

$$\frac{dx_2}{dt} = q_2(t) - \frac{s_2}{c}(1 - L) + \frac{s_2}{s_1}u, \tag{51}$$

where we have introduced a *control variable u* defined by

$$u = \frac{s_1 g_1}{c}. \tag{52}$$

We seek to minimize the delay to the users, which is given by

$$x_0 = \int_0^T (x_1 + x_2)\,dt, \tag{53}$$

subject to the end conditions

$$x_i(0) = x_i(T) = 0 \qquad i = 1, 2 \tag{54}$$

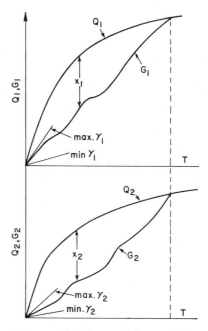

FIGURE 20. The problem of the oversaturated intersection treated by the Pontryagin maximum principle.

and the constraint that the control variable be within the admissible control domain

$$s_1 g_{min}/c = u_{min} \leq u \leq u_{max} = s_1 g_{max}/c. \tag{55}$$

This is a problem in linear control theory, and the linearity leads to a bang–bang solution. The solution, following Pontryagin, proceeds as follows.

Equations 51 and 53 are equivalent to

$$\frac{dx_i}{dt} = f_i(x_j, u) \quad \begin{matrix} i = 0, 1, 2 \\ j = 1, 2 \end{matrix}, \tag{56}$$

where

$$f_0 = x_1 + x_2,$$

$$f_1 = q_1(t) - u, \tag{57}$$

$$f_2 = q_2(t) - s_2\left(1 - \frac{L}{c}\right) + \frac{s_2}{s_1} u.$$

We define the *adjoint variables* ψ_i by

$$\frac{d\psi_i}{dt} = -\sum_{l=0}^{2} \frac{\partial f_l(x_j, u)}{\partial x_i} \psi_l \qquad i = 0, 1, 2. \qquad (58)$$

The system of Eq. 58 can be solved, yielding

$$\psi_0 = k_0,$$
$$\psi_i = -k_0 t + k_i \qquad i = 1, 2, \qquad (59)$$

where k_0, k_1, and k_2 are constants. We then define the Hamiltonian of the system H given by

$$H(\psi_i, x_i, u) = \sum_{l=0}^{2} \psi_l f_l = zu + w, \qquad (60)$$

where

$$z = \left(1 - \frac{s_2}{s_1}\right)k_0 t + \left(-k_1 + \frac{s_2}{s_1}k_2\right),$$

$$w = q_1(-k_0 t + k_1) + \left[q_2 - s_2\left(1 - \frac{L}{c}\right)\right](-k_0 t + k_2) + k_0(x_1 + x_2). \qquad (61)$$

Pontryagin's *maximum principle* in effect states that the optimum control u maximizes the Hamiltonian, subject to appropriate end conditions for the state and adjoint variables. From Eq. 60 we see that H is maximized if u is equal to its maximum value when z is positive and equal to its minimum value when z is negative. Using the signum function of z (sg $z = $ sign of z) we can write the optimum u in the form

$$u = \tfrac{1}{2}[(1 + \text{sg } z)u_{\max} + (1 - \text{sg } z)u_{\min}]. \qquad (62)$$

The remainder of the solution is straightforward, leading to the equations already given for the switch-over time τ and duration of the rush period T.

Complex Oversaturated Systems

A Pair of Intersections

When two oversaturated intersections are close together, they are coupled through the portion of vehicles that pass through both intersections in succession. Consider, for example, the system of two intersections serving streams 1, 2, and 3, which are assumed to go through the intersections (1, 2) and (1, 3) (Figure 21) without turning. Let us further assume that the saturation flows satisfy the relationships

$$s_1 > s_2, \qquad s_1 > s_3, \qquad s_1 < s_2 + s_3. \qquad (63)$$

If we optimize independently the two intersections (1, 2) and (1, 3) according to the preceding discussion of the single intersection, we find that traffic

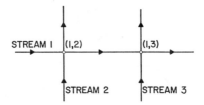

FIGURE 21. A pair of oversaturated
intersections in tandem.

stream 1 must receive preferential treatment. If, however, we take the point
of view that streams 2 and 3 are like traffic streams on a boulevard intersecting
stream 1, then on the strength of the last of Eqs. 63 it is stream $(2 + 3)$ which
must receive preferential treatment. It is this second point of view that must
be taken, because if one treats each intersection independently, he counts
the delay to stream 1 roughly twice, once at intersection $(1, 2)$ and once at
intersection $(1, 3)$.

Let us now assume that the transit time and queueing storage between
intersections are negligible and that the intersections become oversaturated
roughly at the same time. We can obtain the optimum strategy of operation
of the two lights by applying a policy improvement scheme such as the one
used for the single intersection, namely, trading off delays of the various
streams to reduce the overall delay. We begin by obtaining the earliest end
of the rush period for each one of the two intersections, considered as
isolated. Let these times be T_{12} and T_{13}, respectively, for intersections $(1, 2)$
and $(1, 3)$. We also compute the switch-over times for these intersections,
considered as isolated. These switch-over times τ_{12} and τ_{13} are given by

$$\tau_{1j} = \left[g_{\max} T_{1j} - \left(\frac{c}{s_1} \right) Q_1(T_{1j}) \right] (g_{\max} - g_{\min})^{-1} \qquad j = 2, 3. \qquad (64)$$

If the maximum and minimum service rates for stream 1 are the same at both
intersections, then the relative magnitude of τ_{1j} is the same as that of T_{1j}.
Figure 22 shows the complete solution for the case $T_{12} < T_{13}$, which is
obtained as follows: The service of intersection $(1, 2)$ considered as isolated
is the optimum. No delay trade-off is possible that will reduce the delay of
stream 1, because the corresponding increase of delay for the streams 2 and
3 is greater. However, after the switch-over time τ_{12}, a delay trade-off is
possible between streams 1 and 3 only. The maximum trade-off corresponds
to the exchange of a delay equal to the area of the parallelogram $ABCD$ for
a smaller delay equal to the area $abcd$. This is profitable since it has been
assumed that $s_1 > s_3$. The key to the solution of the problem here is that, in
view of relations Eq. 63 it is profitable to exchange delay of stream 1 for delay
of stream 2 *or* 3, but not both.

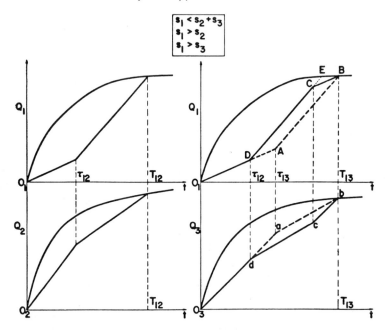

FIGURE 22. Optimum control of a pair of oversaturated intersections.

Thus, the optimal control of the two-intersection system involves, in general, two switch-over times, because there are two control variables in the system, the two splits at the two intersections.

Spill-Back from an Exit Ramp of an Expressway

Consider the system shown in Figure 23, comprising an expressway 1, an exit ramp 2, and the exit highway 3. The intersection of 2 and 3 is assumed to be controlled by a traffic light 4 (as in the case of an observed real situation). It is assumed that this intersection is oversaturated during a rush period. A queue may then build along the exit ramp, and when its length exceeds the storage capacity of the ramp, the queue spills back into the expressway.[36] The spill-back ties up at least one lane, sometimes two, and reduces substantially the throughput of the expressway. The problem is to optimize the operation of the traffic light 4, in order to minimize the delay of the combined streams 1, 2, and 3.

The cumulative demand curves Q_1, Q_2, and Q_3, as well as the maximum and minimum service rates for all the streams are shown in Figure 24. The service rates for streams 2 and 3 are determined directly from the split of the traffic light, with Γ_j being the maximum and γ_j the minimum service rate. The expressway is assumed to have a throughput equal to Γ_1 or γ_1, depending

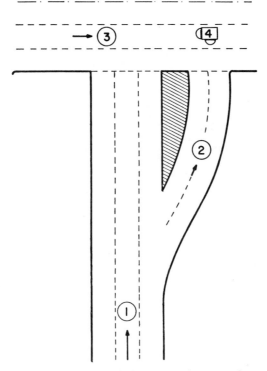

FIGURE 23. Spill-back from an exit ramp of an expressway.

on whether it is unobstructed or clogged due to spill-back. More often than not, the saturation flows s_2 and s_3 are such that

$$s_2 < s_3. \tag{65}$$

According to the discussion in the first part of Section V, the optimum operation of the light 4, considered as isolated, would involve service curves such as O_3EF and o_2ef, with the highway stream 3 receiving preferential treatment. This would produce spill-back from time t_g to time t_h. These times are determined by drawing the curve

$$\bar{Q}_2 = Q_2 - Q_2^*, \tag{66}$$

where Q_2^* is the capacity of the ramp, and finding the intersection points of this curve with O_2ef. Choosing the service curves O_2klmnf and O_3KLMNF can, in this case, eliminate spill-back altogether. The delay to stream 3 is increased by a quantity A_3, the delay to stream 2 is decreased by A_2, and the delay A_1 that would have been caused by spill-back is completely eliminated.

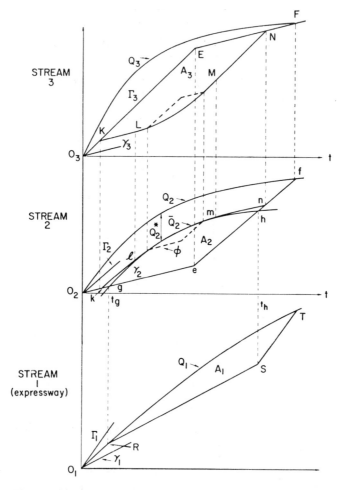

FIGURE 24. Optimum control of the system comprising an expressway, a highway, and an exit ramp from the expressway to the highway.

Since

$$\frac{A_2}{A_3} = \frac{s_2}{s_3},$$ (67)

the total change in the aggregate delay is

$$\delta = A_1 + A_2 - A_3 = A_1 + A_2\left(1 - \frac{s_3}{s_2}\right).$$ (68)

A net reduction of delay results if δ is positive. In this case it pays to adopt the strategy of keeping the queue along the ramp below the critical value Q_2^*. If δ is negative, then spill-back is not as damaging as it appears, at least in terms of total delay, which is minimized by an optimum operation of the traffic light 4, assumed isolated. However, it may still be desirable to prevent congestion on the expressway for safety reasons that may over-ride delay considerations. If this is the case, one may accept a small negative delay trade-off δ.

Assuming that the delay criterion is the dominant one, we find that a critical constant rate q_1 of demand along the expressway exists, which is related to A_2 according to a relationship obtained by setting δ in Eq. 68 equal to zero. Thus,

$$\frac{(\Gamma_1 - \gamma_1)(q_1 - \gamma_1) \tau^2}{\Gamma_1 - q_1} \frac{\tau^2}{2} = A_2\left(\frac{s_3}{s_2} - 1\right), \tag{69}$$

where the left-hand side of Eq. 69 is equal to A_1 and

$$\tau = t_h - t_g. \tag{70}$$

Solving Eq. 69 for q_1, we obtain

$$q_1 = \frac{2A_2(s_3/s_2 - 1)\Gamma_1 + \gamma_1(\Gamma_1 - \gamma_1)\tau^2}{2A_2(s_3/s_2 - 1) + (\Gamma_1 - \gamma_1)\tau^2}. \tag{71}$$

If the demand rate is smaller than q_1, then spill-back is the lesser of two evils, since it corresponds to minimum total delay.

By similar arguments we may investigate the possibility of allowing spill-back during a portion of the interval $(t_h - t_g)$. If the rate of demand along the expressway falls sufficiently below Γ_1, it may be profitable to adopt a strategy such as that corresponding to the dashed line ϕ in the middle diagram of Figure 24 and the complementary service curves for streams 1 and 3 (the last one not shown in Figure 24). This policy will introduce some additional delay to the stream 2, and because of spill-back it will also delay some vehicles in stream 1. It will reduce somewhat the average delay to stream 3. The net change can be computed by an expression similar to Eq. 68, if the exact shapes of Q_2 and Q_1 are known. Finally, if the demand rate along the expressway falls below γ_1, it is always profitable to allow spill-back.

Reversible Lanes

An interesting example of reversible lanes is that of the Lincoln Tunnel between New Jersey and New York City, which becomes oversaturated in both directions during a rush period. One or two out of the six lanes of the tunnel may be reversed during the rush period to balance the service in the two directions. The lane reversal entails a fixed penalty of idle time during the process of reversal of the direction of traffic. A method has been given by this author[37] for finding the assignment of the reversible lane, during the rush period, that minimizes the combined delay to the users of the tunnel who queue up in both directions. The solution is based on the following assumptions:

1. There exists a known time dependence of demand that exceeds the service capability of the tunnel during a rush period.
2. One lane may be reversed, increasing the service rate in one or the other direction by a different amount.
3. Reversal of this lane entails a fixed penalty equivalent to having this lane idle for a fixed period of reversal, E.
4. The throughput of each lane can be maintained at some high constant level as long as there is a queue waiting to be served.

The objective of lane reversal is to minimize delay, subject, perhaps, to some constraints regarding the maximum size of queue which is permissible on the Manhattan side of the tunnel.

The formulation and solution of the problem is shown in Figure 25. The cumulative demand and service (CDS) diagram is plotted for both directions 1 and 2, and for their sum. It differs from the CDS diagram used in the previous cases in that only the excess of cumulative demand and service over the minimum cumulate service is shown. Utilization of the reversible lane increases the service rates by amounts γ_1 and γ_2, and it is assumed that $\gamma_1 > \gamma_2$. Furthermore, we assume that at the end of the rush period the rates of demand are approximately constant. Hence

$$Q_i + b_i t = c_i \qquad i = 1, 2, \tag{72}$$

where Q_i are the cumulative demands, t is the time, and b_i, c_i are constants. If we wish to exhaust both queues as early as possible, then the optimum policy corresponds to the service curves $O_1 A_1 B_1$ and $O_2 A_2 B_2$, or $OAA'B$ in the sum diagram. The end of the rush hour, T, and the time of lane reversal, τ_1, are given by

$$T = \frac{c_1 \gamma_2 + c_2 \gamma_1 + \gamma_1 \gamma_2 E}{\gamma_1 b_2 + \gamma_2 b_1 + \gamma_1 \gamma_2},$$

$$\tau_1 = \frac{c_1 b_2 - c_2 b_1 + \gamma_2 (c_1 - b_1 E)}{\gamma_1 b_2 + \gamma_2 b_1 + \gamma_1 \gamma_2}. \tag{73}$$

If we must start the rush period with the reversible lane assigned to stream 2, we can still exhaust both queues at T by reversing the lane at time τ_2 given by the second of Eqs. 73, but with the indices 1 and 2 reversed. In comparison

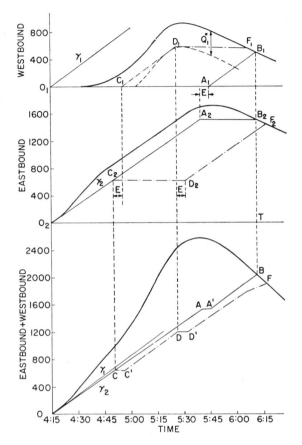

FIGURE 25. Optimum assignment of a reversible lane in an oversaturated traffic link.

with the optimal policy, this second policy involves some additional delay equal to the area $OAA'BC'CO$.

There is also the possibility of a further reduction of the aggregate delay, if one is willing to extend the rush period for one of the streams. The total delay has a stationary value if we adopt a policy given by the service curves $O_1D_1F_1$ and $O_2D_2F_2$. This policy is obtained by extending the assignment of the reversible lane to stream 1 past the time τ_1, so that the second legs of the service curves have durations which satisfy the relationship

$$\frac{(D_1F_1)}{(D_2F_2)} = \frac{\gamma_2}{\gamma_1}.$$ (74)

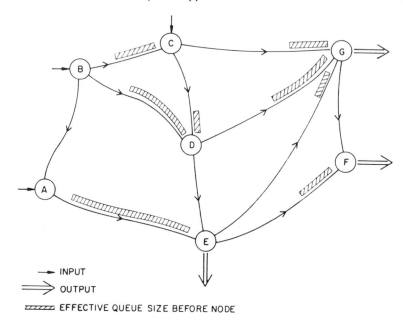

FIGURE 26. A schematic diagram of a traffic store-and-foreward network.

Generalization to Complex Congested Transportation Systems

The preceding examples are special cases of congested, store-and-forward networks, such as the one depicted in Figure 26, which are characterized by the following requirements and constraints:

1. There exist some time-dependent, possibly stochastic, origin–destination requirements of traffic volumes, which cannot be fully accommodated during a certain rush period.
2. With each arc of the network are associated three parameters:
   (a) A (fixed) travel time, or travel cost,
   (b) A capacity restraint, the maximum volume of traffic which can move along the arc per unit of time,
   (c) A storage capability allowing traffic to be stored just before a node at the end of an arc, and be dispatched when roadway space becomes available.
3. Traffic moves at constant speed from node to node, and then either moves past the node or is stored there. (In practice, of course, the storage space is distributed along an arc, and storage time is accrued in installments along the arc.)

Many transportation networks have the above general features. Examples are congested road networks,* and a congested system of airports and the airspace where stacking, holding patterns, and other storage activities take place. In optimizing the operation of such systems, the primary objective would be to minimize a delay function, possibly weighted for priority assignments, subject to operational and capacity constraints. An optimization procedure would entail two main tasks:

1. A *route assignment* for every traffic unit, the assignment taking into account the current load on each section of the network.
2. *Switching* at the nodes, allocating the roadway usage to various traffic units.

There is no complete methodology for the solution of this most general optimization problem. If, however, the route assignment is stipulated, then the switching problem may be cast into the framework of control theory. In this formulation, the state variables are the queues before the nodes which have originated from upstream nodes and are waiting to be dispatched to various downstream nodes of the network. Their rate of change depends on some given traffic input rates and on some control variables that represent the switching at the nodes. The objective is to minimize the delay, which is a function of the state variables, subject to some appropriate constraints on the control variables. The mathematical formulation of the problem is the following. Minimize

$$D = \sum_j \int_0^T \alpha_j Q_j \, dt \qquad j = 1, \ldots, J \tag{75}$$

given that

$$\frac{dQ_j}{dt} = f_j(I_k, u_l), \qquad \begin{matrix} k = 1, \ldots, K, \\ l = 1, \ldots, L, \end{matrix} \tag{76}$$

and subject to some end conditions

$$Q_j(0) \text{ and } Q_j(T) \text{ given} \tag{77}$$

and constraints on the control variables u_l:

$$u_l \in U. \tag{78}$$

In the above equations, Q_j are the state variables, α_j are weighting constraints, I_k the given traffic inputs (with known destinations and route assignments), u_l the control variables, U a control domain, and f_j some functionals which

---

* In a congested road network, an arc is not necessarily a section between two intersections. It may be convenient to define as nodes only major bottlenecks or major route-switching points, in which case arcs may contain several intersections.

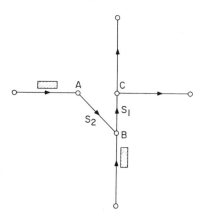

FIGURE 27. The oversaturated intersection viewed as a store-and-forward network.

may possibly involve not only the instantaneous values of $I_k(t)$ and $u_k(t)$ but also their time history.

One possible general method for the solution of the above problem is through reduction to a mathematical programming problem by means of discretization, that is, by replacing the time derivatives by differences and the integral of Eq. 75 by a sum. If, furthermore, the functions f_j are linear in u_l, discretization yields a linear programming problem. We shall illustrate this procedure by considering once more the case of the single oversaturated intersection. Referring to Figure 27 we observe that this case can be viewed as a special case of a store-and-forward network of a very simple "tree" configuration. In Figure 27 the arcs AB and BC are dummy arcs associated with zero travel time and storage space, but finite capacities s_1 and s_2, respectively. Switching takes place at B, and storage just before nodes A and B.

Let us assume that the cumulative demand curves are given by*

$$Q_i = A_i + B_i t \qquad i = 1, 2. \tag{79}$$

as shown in Figure 28. Rather than discussing an example of an arbitrary subdivision of the rush period, we shall assume that this period of control is divided into two periods, and formulate the problem as a parametric linear programming problem with parameters τ, the switch-over time, and T,

---

* It has already been pointed out that the asymptotic behavior of Q_j is sufficient for optimization purposes.

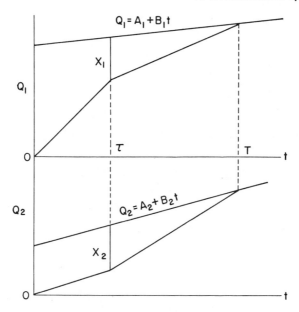

FIGURE 28. Linear programming solution of the problem of the oversaturated intersection.

the duration of the rush period. Referring to Figure 28, we seek to minimize the delay D given by

$$D = \tfrac{1}{2}\{(A_1 + A_2)\tau + (x_1 + x_2)T\}, \tag{80}$$

where x_1 and x_2 are two independent variables of the problem. The constraints of the problem, already discussed in the beginning of Section V are that the rates of service for both traffic streams, given by the slopes of the cumulative service curves, must lie within two bounds. Using these constraints we find that the x_i must satisfy the inequalities

$$\frac{g_{\max}}{c} \geq \frac{A_i + B_i\tau - x_i}{s_i\tau} \geq \frac{g_{\min}}{c}$$

$$\frac{g_{\max}}{c} \geq \frac{x_i + B_i(T - \tau)}{s_i(T - \tau)} \geq \frac{g_{\min}}{c} \qquad i = 1, 2, \tag{81}$$

where $g_{\max}$ and $g_{\min}$ are the maximum and minimum green phase and c is the cycle. In addition, the green phases required to serve the two streams must be at most equal to the light cycle minus the lost time L, a condition which yields the inequalities

$$\frac{A_1 + B_1\tau - x_1}{s_1\tau} + \frac{A_2 + B_2\tau - x_2}{s_2\tau} \le 1 - \frac{L}{c},$$

$$\frac{x_1 + B_1(T - \tau)}{s_1(T - \tau)} + \frac{x_2 + B_2(T - \tau)}{s_2(T - \tau)} \le 1 - \frac{L}{c}. \tag{82}$$

The constraints, Eqs. 81 and 82, can be rewritten in the form

$$\frac{A_i + B_i\tau}{s_i} - \frac{\tau g_{min}}{c} \ge \frac{x_i}{s_i} \ge \frac{A_i + B_i\tau}{s_i} - \frac{\tau g_{max}}{c}, \tag{83a}$$

$$\frac{(T - \tau)(g_{max} - B_i)}{c} \ge \frac{x_i}{s_i} \ge \frac{(T - \tau)(g_{min} - B_i)}{c}, \tag{83b}$$

$$\frac{x_1}{s_1} + \frac{x_2}{s_2} \ge \frac{A_1 + B_1\tau}{s_1} + \frac{A_2 + B_2\tau}{s_2} - \tau\left(1 - \frac{L}{c}\right), \tag{83c}$$

$$\frac{x_1}{s_1} + \frac{x_2}{s_2} \le (T - \tau)\left(1 - \frac{L}{c}\right) - \left(\frac{B_1}{s_1} + \frac{B_2}{s_2}\right), \tag{83d}$$

The objective of minimizing D, given by Eq. 77, and the constraints given by Eq. 83 comprise the statement of the problem. It can be recognized as a parametric linear programming problem, where the two parameters T and τ must be so selected as to ensure feasibility and optimality of a solution. The solution can be obtained as follows:

Each one of the Eqs. 83a and 83b defines a rectangular region in state space x_1, x_2 (Figure 29). Equations 83c and 83d, taken with the equality sign are two parallel lines, the diagonals from the upper left-hand corner to the lower right-hand corner of these rectangles. In order for a feasible solution to exist, the diagonal CD corresponding to Eq. 83d must be above or coincidental with the diagonal AB corresponding to Eq. 83c. This condition for feasibility yields a condition for T, namely,

$$T\left(1 - \frac{L}{c}\right) - \frac{A_1 + B_1T}{s_1} - \frac{A_2 + B_2T}{s_2} \ge 0. \tag{89}$$

Equation 84 expresses the fact that the total available green time, $T[1 - (L/c)]$, must be equal to or greater than the sum of the green times required to serve the total number of vehicles that demand service during the time between O and T, along both directions 1 and 2.

For any value of T satisfying Eq. 84 there exists a linear programming solution. It is intuitively obvious that the best value of T is the one satisfying

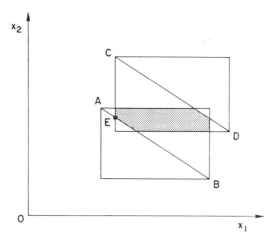

FIGURE 29. Feasibility domain and solution of the LP problem for the oversaturated intersection.

equality 84, since it corresponds to maximum utilization of green time. The remainder of the solution can be obtained using a standard linear programming technique. The variable τ may be treated as a third-state variable without destroying the linearity of the objective function and constraints.

As is well known, linear programming solutions correspond to an apex of the feasibility domain; that is, they satisfy some constraints with the signs of equality. This observation provides an instructive justification of the bang–bang nature of the optimal control for a problem in linear control theory.

VI. CONTROL OF CRITICAL TRAFFIC LINKS

Many expressways appear to be misnamed, at least during periods of rush traffic. The reason is generally one of improper allocation of traffic facilities; the cumulative net input from several ramps exceeds the throughput of the expressway at some point that then becomes a bottleneck of the system. Once congestion sets in, we may also observe what we might call the "revolving door effect." The throughput of the bottleneck is lower during congestion than it is when traffic is moving freely (Figure 30). This degradation is particularly evident in certain facilities, such as the tunnels under the Hudson River going into New York City, where the geometric configuration tends to induce a strong asymmetry between the acceleration and deceleration capability of individual vehicles. This asymmetry produces larger time headways downstream from a point of deceleration than upstream (Figure 31), and hence a loss of throughput of the facility. Even in the absence of

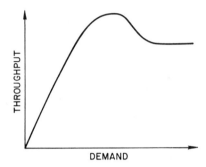

FIGURE 30. Reduction of throughput due to congestion observed at
the Lincoln Tunnel.

such degradation, however, it is desirable to prevent congestion in expressways, if only for reasons of safety.

Over the past decade, there has been much experimentation in freeway
control in an attempt to develop a methodology for preventing congestion
and for operating critical traffic facilities at peak efficiency. The control of
freeways involves controlling the traffic input at a number of ramps, or even
closing these ramps altogether during peak periods. In the case of the tunnels

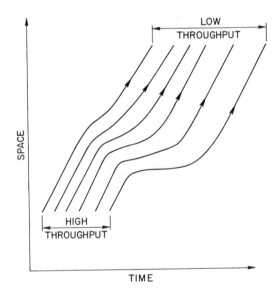

FIGURE 31. Shockwaves that cause reduction
of throughput in tunnels during congestion.

into New York City, the objective has been to keep them operating at peak efficiency by preventing congestion which decreases throughput. We shall give an overview of these two types of control, followed by a discussion of filtering techniques of control theory for the evaluation of traffic densities in critical traffic links.

Control of Expressways

There is no disagreement concerning the need for input control of freeway ramps. However, there are different philosophies concerning this control, some of which have been implemented during various freeway control experiments. The two earliest projects on freeway control were the Eisenhower Expressway project in Chicago[38] and the John Lodge Expressway in Detroit.[39] In their early stages the two projects were substantially different in orientation. The Eisenhower Expressway project emphasized attempts to quantify the description of traffic, collect data, and develop an automatic control system. The John Lodge project emphasized closed-circuit television surveillance and manual intervention of an emergency nature, such as the closing of an entrance ramp, or the dispatching of a patrol car to a scene of trouble. The current thinking in both these projects and in other ones that followed views an expressway and its entrance exit ramps as a tree network, such as that shown with solid lines in Figure 32. If our objective is to prevent congestion on any section of the freeway, we must control the input on one or more of the input ramps during periods of peak demand. Among the various philosophies of ramp control are the following.

1. The gap acceptance philosophy promulgated by Drew.[40] According to this philosophy, cars are allowed to enter a freeway only when there is a sufficiently long gap in the lane nearest to the entrance ramp. One danger in this policy is that the large gap may be compensated for by a high-density wave just in front, in which case the availability of space for the entering car is only illusory.
2. The Wattleworth scheme[41] of maximizing service while satisfying constraints on the various sections of the freeway. This scheme assumes some known, constant origin–destination requirements of all cars entering the freeway at various entrance ramps. It then allows rates of input such that the capacity constraints are satisfied and the overall number of vehicles served per unit of time is maximized. It may be proved that this formulation leads to a linear programming problem that can be solved by standard algorithms, such as the simplex method. One basic defect of the method is that it does not take into account the variation of demands with time. Wattleworth suggests a partial correction of this deficiency by subdividing the peak period into several periods associated with different values of the

various parameters of the problem, such as demands and origin–destination coefficients. Another defect is that the control is effectively an open-loop one, insensitive to the current densities in the various sections of the freeway.

3. Other approaches to freeway ramp control are oriented toward balancing the sizes of queues at the entrance ramps. Also, some thought has been given to considering the surface streets together with the freeway as an integrated traffic system, but little work has been done along this line so far. The reader will recognize the freeway control problem as another case of the facilities allocation problem discussed in Section V. Thus, the network of Figure 32, extended to include surface streets, which are shown with dashed lines, is a store-and-forward network. Flow in this network is controlled at the input ramps and possibly along the approaches to those ramps. In principle, one can obtain the optimal control of this network for any given inputs and origin–destination specifications, as discussed in Section V. However, it must be emphasized that any open-loop operation on the basis of an optimum design obtained off-line using average inputs has very little chance of success. There are enough perturbations in freeway traffic that can upset the schedule of allocation of facilities on the basis of average inputs. Experience has shown that ramp control, or the overall control of the system of Figure 32, must respond to observed

FIGURE 32. Representation of a freeway and its neighboring surface streets as a network. The solid lines represent the freeway and its ramps, and form a tree network.

conditions *along* the freeway. It is in this kind of response that many present systems leave room for improvement. Both our ability to measure accurately traffic conditions over the entire length of a freeway, and our knowledge of the proper control action for any given traffic conditions, are still relatively limited.

The preceding discussion of freeway control is by no means exhaustive, particularly with regard to experimentation, which has been quite extensive over the last ten years. This experimentation has laid the groundwork for sophisticated freeway control systems through development and testing of necessary hardware such as ramp-metering devices, detection hardware, and so on.

Control of Traffic at the Lincoln Tunnel

Experiments, starting in 1959, showed that the throughput of a tunnel could be increased by regulating the input so as to prevent congestion.[42] The first experiments involved open-loop control, namely, metering the traffic to allow a fixed number of cars, usually 22 per lane, to enter the tunnel every minute. This fixed number corresponded to the maximum throughput rate that had been observed at the tunnel for any sustained period. As might be expected, this control worked well only part of the time. However, once the tunnel became congested, for any reason, the open-loop control of the input could not bring it back to optimum operating level. Closed-loop experiments were initiated, first using fixed-logic control hardware,[43] and eventually an on-line digital computer.[44] The use of a computer was first tested during a joint study of The Port of New York Authority and IBM that ran from 1966 until 1969 and aimed at developing a methodology for the surveillance and control of the tunnel by an on-line computer.

The tunnel was divided into three sections, each approximately one-half mile long, by four observation points, or traps, where pairs of photocells about 14 ft apart were used to detect passing vehicles (Figure 33). By processing the signals from these detectors with a computer, the speeds and lengths of passing vehicles were determined. Then, matching of the patterns of lengths at consecutive traps provided identification of individual cars and hence yielded information about the counts K_{12}, K_{23}, and K_{34} in the three sections of the tunnel.

The control strategy was developed by analyzing data of past performance of the tunnel. The algorithm was based on the premise that the performance of the tunnel can be predicted reasonably well by a function of three state variables, the three vehicle counts in the three sections between traps, for each lane. Measurements during uncontrolled conditions indicated that the premise was correct, in that the probability of congestion, in the absence of

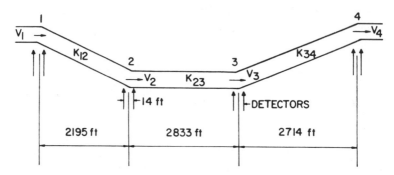

FIGURE 33. Geometric configuration of the south tube Lincoln Tunnel (not in scale) showing the four locations of the detector pairs.

input control, increased as the three counts K_{12}, K_{23}, and K_{34} were increased. Therefore, a surface was estimated in state space K_{ij} the crossing of which, through increasing values of K_{ij}, triggered the control signal. A second surface, closer to the origin, signaled the removal of the control restraint for decreasing values of K_{ij}. An additional adaptive feature varied these surfaces as a function of the measured speed at the foot of the upgrade that was the most frequently observed bottleneck. This adaptive feature called for earlier application of input control when the observed speed at the foot of the up-grade was unusually low, too low for the observed conditions. It had the overall effect of compensating for unobservable influences, such as conditions outside of the tunnel.

The results of the Lincoln Tunnel experiment indicated that regulation of the input could increase the average throughput by as much as 10%. In addition, the speeds through the tunnel were reasonably high, around 30 miles/hr, instead of the usual 10 miles/hr observed during stop-and-go conditions. This fact, together with the overall reduction of traffic densities, not only was more satisfactory to the drivers but also had the salubrious effect of reduced ventilation requirements for the tunnel and possibly fewer breakdowns of cars due to overheating.

Measurement of Densities

The control of traffic at the Lincoln Tunnel required a fairly accurate estimate of the number of cars in various sections of the tunnel. An estimate of traffic densities would also be a prime requirement of an optimal control system for an oversaturated store-and-forward network such as those discussed in Section V. For this reason we shall conclude this chapter with a discussion of some recent advances in density estimation by application of techniques of control theory.

As was mentioned on page 233, the number of vehicles in the three sections of the Lincoln Tunnel was estimated by identifying individual vehicles through the matching of patterns of lengths of a sequence of vehicles at two consecutive traps. This technique, while very accurate, is somewhat expensive in terms of computer requirements, and it is only practical for an environment where few lane changes take place. However, the availability of the accurate density data for the Lincoln Tunnel made possible the development and testing by Gazis and Knapp[45] of an alternative density estimation algorithm, which is, in principle, applicable to any traffic system.

Let us consider a section of a roadway contained between two detector stations. Let us then assume that we are able to obtain at discrete intervals an estimate of the number of cars in that section, say, z_k at the end of the kth interval. (In the original reference this estimate was obtained by first estimating the travel time of individual vehicles through the section. It was shown later by Szeto and Gazis[47] that other methods for obtaining z_k, for example, a phenomenological relationship between speed and density, can also be used satisfactorily.)

Let y_k represent the true value of the number of vehicles at the end of the kth interval. The "rough" estimate z_k deviates from y_k by an amount u_k assumed to be a random variable with zero mean, that is,

$$z_k = y_k + u_k. \tag{85}$$

Successive values of y_k are related according to the relationship

$$y_k = y_{k-1} + \Delta N_k + w_k, \tag{86}$$

where N_k is the *net input* into the section (input at the entrance minus output at the exit) and w_k is a random measurement error, also assumed to have a zero mean.

The algorithm, based on the technique of Kalman filtering,[46] yields successive values of the estimated number of vehicles in the section $\hat{y}_k$ according to the relationship

$$\hat{y}_k = B_k z_k + (1 - B_k)[y_{k-1} + \Delta N_k], \tag{87}$$

where

$$B_k = \frac{Q + C_{k-1}}{R + Q + C_{k-1}},$$

$$C_k = B_k R, \tag{88}$$

$$Q = E[w_k^2],$$

$$R = E[u_k^2],$$

and E signifies expectation. The algorithm above gives an optimum estimate in the sense that the expectation of the squared error of the estimate is minimized, provided that

$$E[w_k] = E[u_k] = E[w_k u_j] = 0, \tag{89}$$

$$E[w_k w_j] = E[u_k u_j] = 0 \quad \text{if} \quad k \neq j.$$

The process is begun by starting with some values of y_0 and C_0 chosen to satisfy approximately the relationships

$$y_0 = E[y_0], \qquad C_0 = E[\hat{y}_0 - E(y_0)]. \tag{90}$$

Experience with the Lincoln Tunnel data showed that the condition of independence of successive values of u_k ($E[u_k u_j] = 0$) is often violated. In fact it was found that this "autocorrelation function" could be roughly approximated by an exponential function:

$$E[u_k u_j] = R\, d^{k-j} \quad k \geq j. \tag{91}$$

In that case, as was shown in Ref. 45, the preceding algorithm may be modified as follows: Instead of z_k and the B_k, C_k' given in Eq. 91, we use

$$z_k^* = \frac{z_k - dz_{k-1} - d\,\Delta N_k}{1 - d},$$

$$B_k^* = \frac{Q + C_{k-1}}{R^* + Q + C_{k-1}}, \tag{92}$$

$$C_k^* = B_k^* R_k^*,$$

where

$$R^* = E[(y_k - z_k^*)^2] = R\left(\frac{1 + d}{1 - d}\right) + \frac{Qd^2}{(1 - d)^2}. \tag{93}$$

The successive estimates, starting with the initial values y_0 and C_0 of Eq. 90, are given by the relationship

$$\hat{y}_k = B_k^* Z_k^* + (1 - B_k^*)[\hat{y}_{k-1} + \Delta N_k]. \tag{94}$$

Tests of the algorithm using the Lincoln Tunnel data[45] show that, with detection stations about one-half mile apart, one could expect to obtain the number of cars with less than 10% error better than 99% of the time, under any traffic conditions. Such reliability of traffic density estimation is probably more than adequate in most traffic situations.

The preceding algorithm was recently tested[48] using freeway data obtained by aerial photography. The purpose of the test was to find out how accurately the algorithm could estimate densities, assuming different distances between sensors, in a freeway environment where lane changes are frequent. It was found that the density for a single lane was generally obtained with less than

10% error for sensor distances as high as 3000 ft. The density for all lanes combined was obtained with even smaller error, in spite of the fact that this estimate involved some averaging of the speeds over the three lanes of the freeway.

References

1. J. N. Kane, *Famous First Facts*, New York: H. W. Wilson, 3rd ed., 1964, p. 626.
2. B. W. Marsh, "Traffic Control," *Ann. Amer. Soc. Political Social Sci.*, **133**, 90–113 (1927).
3. J. T. Morgan and J. D. C. Little, "Synchronizing Traffic Signals for Maximal Bandwidth," *Operations Res.*, **12**, 896–912 (1964).
4. W. D. Brooks, *Vehicular Traffic Control—Designing Arterial Progression Using a Digital Computer*, Kingston, N.Y.: IBM Data Processing Division, 1964.
5. J. D. C. Little, "The Synchronization of Traffic Signals by Mixed-Integer Linear Programming," *Operations Res.*, **14**, 568–594 (1964).
6. D. C. Gazis, "Traffic Control, Time–Space Diagram and Networks," *Traffic Control—Theory and Instrumentation*, edited by T. R. Horton, New York: Plenum, 1965, pp. 47–63.
7. A. Chang, *IBM J. Res. Develop.*, **11**, No. 4, 436–441 (1967).
8. D. I. Robertson, " 'TRANSYT' Method for Area Traffic Control," *Traffic Engr. & Control*, **11**, 276–281 (1969).
9. *San Jose Traffic Control Project—Final Report*, San Jose, Calif.: IBM Corporation, Data Processing Report, 1966; see also D. C. Gazis and O. Bermant, "Dynamic Traffic Control Systems and the San Jose Experiment," *Proc. 8th Intern. Study Week in Traffic Engineering*, Barcelona, Spain, V (1966).
10. J. A. Hillier, "Appendix to Glasgow's Experiment in Area Traffic Control," *Traffic Engr. & Control*, **7**, 1966.
11. J. A. Hillier and R. S. Lott, "A Method of Linking Traffic Signals to Minimize Delay," *Proc. 8th Intern. Study Week in Traffic Engineering*, Barcelona, Spain, V, 1966.
12. R. E. Allsop, "Choice of Offsets in Linking Traffic Signals," *Traffic Engr. & Control*, **10**, 73–75 (1968).
13. R. E. Allsop, "Selection of Offsets to Minimize Delay to Traffic in a Network Controlled by Fixed-Time Signals," *Transport. Sci.*, **2**, 1–13 (1968).
14. "SIGOP," Traffic Research Corporation, New York, 1966. Distributed by Clearinghouse for Federal Scientific and Technical Information, PB 17 37 38.
15. D. W. Ross, "Traffic Control and Highway Networks," *Networks*, **2**, 97–123 (1972).
16. D. J. Buckley, L. G. Hackett, D. J. K. Keuneman, and L. A. Beranek, "Optimum Timing for Coordinated Traffic Signals," *Proc. Aust. Road Res. Board*, **3**, Pt. 1, 334–353 (1966).

17. N. Gartner, "Constraining Relations among Offsets in Synchronized Signal Networks," Letter to the Editor, *Transport. Sci.*, **6**, 88–93 (1972).

18. E. Bavarez and G. F. Newell, "Traffic Signal Synchronization on a One-Way Street," *Transport. Sci.*, **1**, 55–73 (1967).

19. I. Okutani, "Synchronization of Traffic Signals in a Network for Loss Minimizing Offsets," *Proc. 5th Intern. Symp. on the Theory of Traffic Flow and Transportation*, edited by G. F. Newell, New York: Elsevier, 1972, pp. 297–312.

20. N. Gartner, "Optimal Synchronization of Traffic Signal Networks by Dynamic Programming," Ref. 19, pp. 281–295.

21. R. E. Bellman and S. E. Dreyfus, *Applied Dynamic Programming*, Princeton, N.J.: Princeton University Press, 1962.

22. S. Katz, "A Discrete Version of Pontryagin's Maximum Principle," *J. Electron. Control*, **13**, No. 2, 8 (1962).

23. J. Holroyd and J. A. Hillier, "Area Traffic Control in Glasgow—A Summary of Results from Four Control Schemes," *Traffic Engr. & Control*, **11**, 220–223, 1969.

24. M. C. Dunne and R. B. Potts, "Algorithm for Traffic Control," *Operations Res.*, **12**, 870–881, 1964.

25. R. B. Grafton and G. F. Newell, "Optimal Policies for the Control of an Undersaturated Intersection," *Proc. 3rd Intern. Symp. on Traffic Theory*, edited by L. C. Edie, R. Herman, and R. W. Rothery, New York: Elsevier, 1967, pp. 239–257.

26. J. D. van Zijverden and H. Kwakernaak, "A New Approach to Traffic-Actuated Computer Control of Intersections," *Proc. 4th Intern. Symp. on Traffic Theory*, edited by W. Leutzbach and P. Baron, Bonn: Bundesminister fur Verkehr, 1919, pp. 113–117.

27. H. Kwakernaak, "On-Line Dynamic Optimization of Stochastic Control Systems," *Proc. 3rd Intern. Cong. of Intern. Federation of Automatic Control*, London, June 1966.

28. A. J. Miller, "A Computer Control System for Traffic Networks," *Proc. 2nd Intern. Symp. on Traffic Theory*, edited by J. Almond, Paris: Organization for Economic Cooperation and Development, 1965, pp. 200–220.

29. D. W. Ross, R. C. Sandys, and J. L. Schlaefli, *A Computer Control Scheme for Critical-Intersection Control in an Urban Network*, Menlo Park, Calif.: Stanford Research Institute, 1970.

30. M. Koshi, "On-Line Feedback Control of Offsets for Area Control of Traffic," Ref. 19, pp. 269–280.

31. D. C. Gazis and R. B. Potts, "Route Control at Critical Intersections," *Proc. Aust. Road Res. Board*, **3**, Pt. 1, 354–363 (1966).

32. F. V. Webster, "Traffic Signal Settings," Road Research Technical Paper No. 39, Road Research Laboratory, England, 1958.

33. D. C. Gazis and R. B. Potts, "The Over-Saturated Intersection," *Proc. 2nd Intern. Symp. on the Theory of Road Traffic Flow*, edited by J. Almond, Paris: Organization for Economic Cooperation and Development, 1965, pp. 221–237.

34. D. C. Gazis, "Optimum Control of a System of Oversaturated Intersections," *Operations Res.*, **12**, 815–831 (1964).

35. L. S. Pontryagin, V. G. Boltyanskii, R. V. Gamkrelidze, and E. F. Mishchenko, *The Mathematical Theory of Optimal Processes*, translated by K. N. Trirogoff, New York: Wiley-Interscience, 1962.

36. D. C. Gazis, "Spillback from an Exit Ramp of an Expressway," *Highway Res. Rec.*, **89,** pp. 39–46 (1965).

37. D. C. Gazis, "Optimum Assignment of a Reversible Lane in an Oversaturated Traffic Link," Ref. 25, pp. 181–190.

38. A. D. May, "Experimentation with Manual and Automatic Ramp Control," *Highway Res. Rec.*, **59,** 9–38 (1964).

39. E. F. Gervais, "Optimization of Freeway Traffic by Ramp Control," *Highway Res. Rec.*, **59,** 104 (1964).

40. D. R. Drew, "A Study of Freeway Traffic Congestion," Ph.D. dissertation, Texas A & M University, 1964.

41. J. A. Wattleworth and D. S. Berry, "Peak-Period Control of a Freeway System—Some Theoretical Investigations," *Highway Res. Rec.*, **89,** 1–25 (1965).

42. H. Greenberg and A. Daow, "The Control of Traffic Flow to Increase the Flow," *Operations Res.*, **8,** 524–532 (1960).

43. R. S. Foote and K. W. Crowley, "Developing Density Controls for Improved Traffic Operations," *Highway Res. Rec.*, **154,** No. 1430 (1967).

44. D. C. Gazis and R. S. Foote, "Surveillance and Control of Tunnel Traffic by an On-Line Digital Computer," *Transport. Sci.*, **3,** 255–275 (1969).

45. D. C. Gazis and C. H. Knapp, "On-Line Estimation of Traffic Densities from Time-Series of Flow and Speed Data," *Transport. Sci.*, **5,** 283–301 (1971).

46. R. E. Kalman, "A New Approach to Linear Filtering and Prediction Problems," *Trans. ASME (Series D., J. Basic Eng.)*, **82,** 34–45 (1960).

47. M. W. Szeto and D. C. Gazis, "Application of Kalman Filtering to the Surveillance and Control of Traffic Systems," *Transport. Sci.*, **6,** 419–439 (1972).

48. D. C. Gazis and M. W. Szeto, "On the Design of Density Measuring Systems for Freeways," *Highway Res. Rec.* (to be published).

CHAPTER 4

Traffic Generation, Distribution, and Assignment

Walter Helly

+++

Contents

I. Networks—A Descriptive Framework   242

    Introduction   242
    Macroscopic Framework   243
    Detailed Structure   244
       The Graph of a Network   244
       Network Flow   245
       Practical Labeling Strategy   246

II. Models for Predicting or Synthesizing Network Occupancy   248

    The Logical Progression: Generation, Distribution, Assignment   248
    Trip Generation   250
       Prediction of the Predictor Variables   251
       Prediction of the Number of Trips Originating from a Zone   259
    Trip Distribution   260
       Growth Factor Models   262
       Gravity Model   263
       Intervening Opportunity Model   264
       Minimum Entropy Formulation   265
    Modal Split and Diversion Assignment   266
       Modal Split   266
       Diversion Assignment   268
    Assignment   269
       All-or-Nothing Viewpoint   269
       All-or-Nothing Assignment without Capacity Restraint   270
       All-or-Nothing Assignment with Capacity Restraint   274
       Multiple Route Assignment   275
       Smeed's Land Use Accounting Model   276

III. Facility Planning and Temporal Control   279

    Viewpoint 279
    Network Performance Measures 279
    Facility Planning 282
    Facility Management 283

References   285

I. NETWORKS—A DESCRIPTIVE FRAMEWORK

Introduction

One studies vehicular traffic networks to predict usage, to plan facilities, and to control the facilities so as to meet objective efficiency criteria. To achieve these ends, it is necessary to describe the networks, the territories in which they are embedded, and their users in a manner suitable for analysis and simulation.

Because the road network of any urban area is quite complex, one is tempted to try macroscopic analysis and to carry it on in a gross manner as far as possible. Thus an area of interest might be divided into relatively few zones, with all of a zone's properties assumed to be uniform throughout its area. One may even collapse a zone conceptually and associate with it a single geographic *centroid* point. Similarly, the road network may be idealized into a rudimentary *spider web* interconnecting the zone centroids. The ultimate simplification is to ignore roads as such altogether, replacing them by an equivalent amount of land area devoted to movement. One may also suppress all time dependence in the actual traffic flows and thus assume a never-changing average network occupancy. Although we shall exhibit work based on such gross simplifications, specific planning or control problems usually require more detail.

At the other extreme would be a time-dependent microscopic analysis wherein one includes explicitly every user of the network, every vehicle, and every road. Since the users' activities in time and location are not precisely the same every day, some probabilistic assumptions would be required. However, except for these, precise microscopic simulation is entirely possible in principle. In practice, however, a system of as few as four or five inter-sections, the streets connecting them, and the trip end locations abutting on those streets together already represent an enormous data collection and computer programming challenge. The simulation methodology is the lesser problem. The procurement of inputs and parameters for a full network description of sufficient accuracy to warrant such a full description is a nearly impossible task.

For a realistic and practical network analysis, one usually is forced to choose a compromise approach. Land use and occupancy, the consequent generation of trips, and the distribution of these trips among possible destinations are consolidated into macroscopic zonal averages, with no effort made to locate or label individual travelers. However the assignment of traffic to the road network, in the sense of indicating how many vehicles use each road, is done in a more detailed manner because a major objective in network analysis is to draw conclusions on the performance of specific roads, intersections, or control schemes. Still one simplifies as much as possible by simulating only those roads which carry substantial traffic volumes and by either ignoring time dependence or by replacing real, continuous variations in time with very coarse discrete analogs.

Macroscopic Framework

A network area of interest usually is divided into a number of zones. Ideally a zone should have a quite uniform land use and should be devoted entirely to one function, such as single family houses or retail commerce. The zone should be sufficiently compact so that one centroid point can adequately represent all of that zone's location for travel times to the rest of the network. Thus the ideal zone is very small, and a vast number would be required to span a city. Because of budget and computer memory limitations, it is inevitable that somewhat fewer than an ideal number of zones are delineated and that some zones encompass fairly large areas or varied land uses. Even with standards thus quite limited, a large city may easily require more than a thousand zones.

To bound the study network, one introduces a formal boundary within which all the territory is to be handled with uniform detail on a zonal basis. The world outside the boundary is treated formally as one or more zones, but no effort is made to simulate traffic routing there. Thus the traffic to and from the outside world is effectively assumed to terminate at the boundary. To simplify matters, it is desirable to select a boundary with as few major road crossings as possible.

A zone is described by an inventory of its population, land use, vehicles, and transport facilities. Such an inventory is obtained from census data, tax maps, vehicle registrations, and special surveys.

The travel volumes and characteristics within a zone and between zones can be synthesized approximately by use of the inventory above supplemented by counts of vehicles passing road control points or entering employment, shopping, and similar areas. Such a synthesis is likely to be quite inaccurate and suitable only for the grossest analyses. The proper, though expensive, approach is to interview personally or by questionnaire a substantial sample of the population. The aim of such a survey is to obtain the number of trips, classified by purpose and mode of travel, from each zone to itself and to every other zone of the network. Survey data can be checked and refined by comparison of the trip volumes obtained therefrom with corresponding trip volumes obtained by counting vehicles. To do this, all vehicles crossing a *screen line*, which divides the network into two parts, or *cordon*, enclosing part of it, are counted for a statistically satisfactory time period.

All trips during a specified time, such as a 24-hr weekday, using a specified mode, such as private automobiles, for a specified purpose, such as shopping, can conveniently be summarized by a square matrix *trip table*. Each entry t_{ij} of the matrix represents the number of trips from zone i to zone j.

Detailed Structure

The Graph of a Network

A formal, unambiguous mathematical description of a physical network's routes is called a *graph* of that network. The basic elements of a network graph are *links* and *nodes*. A link usually represents some physical distance (or time) that may be traversed on the street network. A complete description of a link would provide travel time and/or cost as functions of link occupancy, rate of flow, time, and other pertinent factors. A node represents a terminus of one or more links. A node may be used to (1) join the termini of a number of links, or (2) represent a *source* (= origin) or *sink* (= destination) of traffic. The traffic originating or terminating in a network zone is assigned to one or more appropriately located nodes of the network graph.

Let the nodes of the network be labeled 1, 2, . . . , N. It will be assumed that all links are *directed links*, meaning that they can be traversed in only one direction. A directed link from node i to node j is represented by (i, j). If a network segment between nodes i and j carries traffic in both directions, it is represented by two links, (i, j) and (j, i). Furthermore, it will be assumed that there is at most one link from any one node to any other node and that there are no links from any one node to itself. Thus (i, j) is unique and (i, i) is not allowed. These assumptions impose no difficulties in the analysis of human travel.

This labeling strategy enables one to describe a graph by listing all the links of that graph. For example, the network of Figure 1 is described by (1, 2), (2, 1), (2, 3), (2, 5), (3, 2), (3, 4), (3, 6), (4, 5), (5, 6), and (6, 3).

A *chain* from node i to node j is a specified route of one or more links all directed the same way, thus permitting travel from i to j. A *cycle* is a chain that terminates at its origin node. A *path* between nodes i and j is a route of links, not all of which are necessarily directed the same way. In Figure 1, (1, 2)(2, 3)(3, 6),   (1, 2)(2, 5)(5, 6),   and   (1, 2)(2, 3)(3, 4)(4, 5)(5, 6)   are

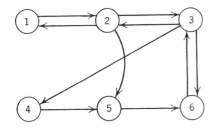

FIGURE 1. A network graph illustration.

chains from node 1 to node 6. (2, 5)(5, 6)(6, 3)(3, 2) is a cycle from node 2 to itself. A path from node 4 to node 1, which is not a chain, is (4, 5)(2, 5)(2, 1). This is also a path from node 1 to node 4.

Node j is *accessible* to node i if there is a chain from i to j.

A *connected network* is one where there are paths between all node pairs. The network of Figure 1 is connected. A *tree* is all or part of a network which (1) is connected, and (2) contains no cycles. Trees of particular concern in the study of vehicular traffic consist of an origin node and one or more destination nodes, with one and only one chain from the origin to each destination. In Figure 1, the following links constitute a tree of this type from node 1 to all other nodes: (1, 2), (2, 5), (5, 6), (6, 3), and (3, 4).

The discussion above on network graph terminology is limited to the immediate requirements of this book. A far more extensive language has been developed. (See Ref. 1 for the mathematician's view, and Ref. 2 for more detailed applications to vehicular traffic.)

Network Flow

A network graph, as exhibited in the previous paragraph, represents the connectivity of the road system. It remains to embellish the graph with a descriptive methodology to represent how the real network serves traffic. The basic approach is to associate a cost c_{ij} with every link (i, j) of the network. Normally one does not assign costs to the nodes.

The simplest view to take for a link cost is to make it the travel time from the origin node to the destination node. This approach is used when it is assumed that travel time is the only matter of interest to the road user when he selects his route. If other factors, such as distance or cost, enter significantly into the user's choice of route, c_{ij} may be made an appropriate function of the several pertinent variables.

Real road links have limited capacity in the sense that the greater the flow in vehicles per unit time, the longer is the average time taken by a vehicle to traverse a link. Thus the link cost may well be made a function of the link usage.

Each vehicle is assumed to enter and to depart the network at specified nodes. The particular chain of links, selected by a vehicle to traverse the network between origin and destination, will be one which minimizes some sort of cost function. The most obvious function, and the one generally used, is the sum of the link costs for all links traversed between origin and destination.

An *assignment* of traffic to the network is a selection of network routes for all vehicles traveling during a specified time interval. If one sums the vehicles using a given link, the usage of that link is obtained. Assignment may be made at various levels of sophistication:

1. *Time-independent assignment.* One assumes that all traffic assigned to a link for a study interval will face the same cost for that link. This cost may be dependent on the number of vehicles assigned. The study interval may be a day, in which case the time independence assumption is most crude. More reasonably, it may be a single specified hour of the day.
2. *Semi-time-dependent assignment.* The study interval is divided into fairly long subintervals, each perhaps 5–30 min long. All traffic originating during one subinterval is assumed to face a time-independent common road cost picture. However cost and consequent usage of a link may vary from one subinterval to the next.
3. *Fully time-dependent assignment.* Time is divided into short intervals that are viewed to be vanishingly small, at least in principle. Network link costs are reevaluated for each interval. Vehicles normally do not start and complete their trips within one time interval. Instead they proceed only as far as appropriate to the link travel times prevailing during that interval.

It often is useful for analysis to state formally a conservation law for the number of vehicles in the network. In general, the law is that every vehicle has both an assigned origin and destination and that it proceeds from origin to destination along a chain of the network graph, using every link in that chain. In other words, no vehicles are created or destroyed in the network except at duly sanctioned sources and sinks.

Practical Labeling Strategy

The most obvious way to graph a network is to select the arterial street system and to discard the purely local access streets, and then to label intersections as nodes with links between to represent the connecting streets. For each zone, an additional node is added, with a zero time link to the principal intersection node in the zone. The additional node is used as the source of trips originating from the zone and as the sink for trips terminating in it. Figure 2 shows how this is done for the immediate vicinity of one intersection.

Though the above approach has been used with some success in practical studies, there are limitations. However, these may be overcome readily by

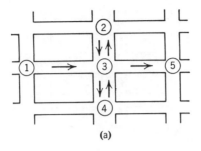

(a)

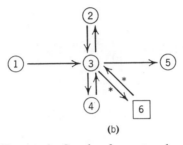

(b)

FIGURE 2. Graph of a network—
nodes at intersections. (a) Part of a
street network. (b) Equivalent part
of network graph. Node 6 has been
added to represent a zone centroid.
Asterisk represents zero time link.

modifications of the labeling strategy. Some of the difficulties and their
solutions are.

1. In a congested environment, delays occur more at intersections than in
   between. Since travel cost or time is associated with the links, it is more
   appropriate to have links pass through intersections and to have the nodes
   at midblock points.
2. Delays are occasioned by turns at intersections. Sometimes turns are
   prohibited entirely. If nodes are placed at intersections, there must be an
   additional superstructure of modeling methodology to cope with turn
   costs or prohibitions. This added complication may be avoided by again
   placing nodes at midblock points and by placing a separate, appropriately
   priced turning link for each allowed turning movement.
3. If all of a zone's trip ends are injected at one node, then there may be
   unrealistic congestion in the vicinity of that node. The solution is to have
   zonal source and sink traffic connected to more than one node. If this
   is done, the single zone centroid node must be replaced by two nodes, one
   for trip originations and the other for terminations. If only one centroid

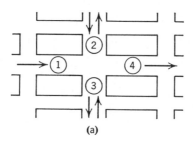

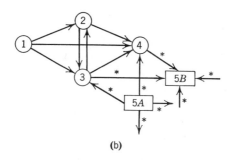

(b)

FIGURE 3. Graph of a network—nodes between intersections. (a) Part of a network. (b) Equivalent part of a network graph. Nodes 5A and 5B have been added to represent a zone centroid. Asterisk represents zero time link.

node is used, the zero-time links to the several network nodes may be used improperly for nonexistent short cuts via the centroid by through traffic that does not have a trip end in the zone.

Figure 3 shows how these improvements are incorporated into the network graph.

II. MODELS FOR PREDICTING OR SYNTHESIZING NETWORK OCCUPANCY

The Logical Progression: Generation, Distribution, Assignment

In planning a road network or any alterations and additions thereto, one needs estimates for the volume and distribution of traffic that will use that network. In the narrowest context of network analysis, the problem might be described as network assignment, the prediction of which routes will be used by specific travelers whose origins and destinations are known. In practice, a much broader analysis is required because one does not know future trip ends and thus must develop appropriate prediction techniques. The overall process consists of three main stages:

1. *Trip generation* is the synthesis of the number of trips originating at all points of the network. If the prediction is exceedingly short-term, perhaps to predict network traffic behavior after some imminent minor change in signalization or control, one may assume that no change occurs between presently surveyed travel habits and those after the change. However, for a longer-term or a dramatic modification, it is necessary to synthesize travel volumes from population, economy, land use, accessibility, and automobile ownership, since any of these may change substantially in at least some of the zones of the network. The number of trips originating in a zone is dependent on these variables, and a trip generation model is an expression of the relationship.
2. *Trip distribution* is the apportionment of trips originating at one place among all possible destinations. The distribution of trips can change most rapidly in response to the opening of a new network link, particularly if the new link is part of a sparse section of the network. Examples of links with dramatic impact include bridges, tunnels, and urban expressway extensions. A trip distribution model is a formal statement of the relationship between (1) the dependent likelihood that a trip terminates at a given destination, and (2) the appropriate predictors, which may include travel time from the trip origin, the number and locations of competing trip ends, and the attractive force of the given destination.
3. *Assignment* is the prediction of the route or routes followed by travelers between a given origin and a given destination. While an individual driver may usually select the shortest time route, this may not always be the case. Furthermore, a practical assignment model should offer a good approximation for the total network usage, which consists of the sum of all trips during a given day or hour. But the route choice of one traveler is influenced by the behavior of others since travel time on a network link is a function of the number of travelers using that link.

Models developed for trip generation, distribution, and assignment must be tested and validated by application to present travel patterns, as deduced from appropriate surveys. Naturally, great caution is appropriate if a model, thus calibrated today, is used to predict activity a generation ahead. There is always the possibility of significant change in human behavior or technology totally unanticipated in the original analysis.

In principle, the travel modeling process can be made fully time-dependent, thus predicting the moments at which trips commence and the locations of all travelers as functions of time. Only if such time dependence is included can one do really accurate network assignment, because the time taken by a traveler to traverse a network segment clearly is dependent on the actual number of other travelers cohabiting his immediate vicinity.

In practice, there rarely is enough information to successfully exhibit a good time-dependent picture even of actual, present day trip generation and distribution. To predict this for the future easily can be quite meaningless. Consequently, the actively used trip generation and distribution models have been made in a time-independent form, designed to present the total sum result of operation for 24 hours or, possibly, for a single rush hour.

There do exist sophisticated time-dependent assignment procedures, generally known as *network simulation models*. Although some of these will be mentioned later in this chapter (see p. 275), it should be made clear that they are used primarily for close simulation of small network arrays, as may be useful for the optimal tuning of traffic control devices. The prediction of overall network operation is likely to remain less ambitious until techniques for obtaining and extrapolating travel patterns are vastly improved.

Trip Generation

When the network travel analysis is to be made for present or very near future traffic, trip generation rates are best obtained by direct traveler surveys, perhaps refined by simple short-term extrapolations. However, when prediction is for a time some years in the future or for a drastically changed land use pattern, then the process of estimating the number of trips originating from each network zone becomes substantially more difficult. There are two stages: (1) the prediction of trip generation prediction variables, such as population, car ownership, shopping opportunities, and number of jobs in each zone of the network; and (2) the prediction of the number of trips originating from a zone, preferably classified by trip purpose, as a function of the predictor variables.

The process is often simplified, for a 24-hr forecast, by assuming that virtually all private travel is in the form of a trip from home to a destination followed, on the same day, by a return trip. Under this assumption, only those private trips originating at residences are generated, but these must be classified by trip purpose. Trips from schools, commercial areas, offices, and factories are made equal in number to the trips terminating at these places. Naturally, errors are introduced by ignoring sequential multidestination travel. Trips by commercial vehicles and by transit must be handled separately.

To estimate the predictor population, economy, and vehicle ownership on a zonal basis, one must first obtain these quantities for the study region as a whole, and then employ a land use prediction model to allocate them among the zones of the network. Only the most basic concepts and models for regional prediction are mentioned here. This is partly because of space limitations and partly because the traffic analyst generally relied on others, such as the Census Bureau, for the appropriate estimates.

Prediction of the Predictor Variables

POPULATION. The number of people residing in a region can be changed only by births, by deaths, and by migration into or out of the region.

In Census Bureau jargon, the "crude birth rate" is the number of births per year per 1000 population. More useful is the "fertility ratio," the number of children under age five per 1000 women of child-bearing age, defined as the range from 15 to 45 years. Both crude birth rates and fertility ratios are not fixed in time or place. The United States had a very low birth rate in the 1930s and a very high one in the 1950s. A developing frontier area starts with few women, usually resulting in a low birth rate and a high fertility rate.

The "crude death rate" is the number of deaths per year per 1000 population. This is determined primarily by the age distribution of the population. In the United States, about two-thirds of all deaths are of persons of age 65 or over. The prediction of future death rates is made by using age-stratified mortality rates rather than the overall crude death rate.

Net migration is the difference between the number of persons who move in and the number who move out of the region. One wishes to predict whether the net migration over a number of years is likely to be inward or outward, the size of the migration, and its age, sex, racial, and economic distribution. Past net migration can be estimated by subtracting the crude natural increase (excess of births over deaths) from the total population change as reported by the census. The leading cause for migration is a disparity between economic opportunities in the study area and in the rest of the country. There has been continued net migration into metropolitan areas from rural areas. This migration has been especially intense since 1950.

The most elementary approach to population forecasting is simple extrapolation, the graphical or mathematical projection of past change. In deducing the shape of the extrapolation curve, it is worth keeping in mind that a stable area is likely to have a relatively constant rate of population change, resulting in an exponential increase (or decrease) with time. A newly developed area, or "new town," is likely to grow largely through in-migration, resulting in an early period of slow population increase, a middle stage of very rapid increase, and a final stable plateau with exponential growth determined by local birth and death rates.

When the study area has a history and prognosis similar to a larger region of which it is a part, and the larger region's future development has already been estimated, it may be appropriate to use the ratio method for population forecasting. The ratio of the study area's population to the larger area's population is calculated for available times in the past. The series of historical ratios is then used as a time series for future extrapolation.

Sometimes it is appropriate to estimate population as a function of future employment. This is particularly the case for an area where employment will change drastically as the result of externally imposed investment. A most common source of such change is the federal government. One must keep in mind that the labor force participation ratio (the ratio of employment to total population) may well be quite abnormal in an area with very rapid employment growth. Furthermore, this ratio appears to be rising constantly, though slowly, for the United States as a whole. It was 0.397 in 1930 and increased to 0.410 by 1960.

Refs. 3 and 4 present further details and extensive bibliographies on population forecasting.

ECONOMY. Economic forecasts are required for employment, for the magnitude and distribution of personal incomes, and for automobile ownership. Rudimentary forecasts may be made by simple extrapolations of past trends or by the ratio method, just as described for population forecasting. The ratio method usually is called the step-down method in this economic context. It has been used for a transportation study of Portland, Oregon.[5] Somewhat more elegant is the economic base multiplier method, used in the Denver transportation study.[6] For this, it is assumed that the local economy depends primarily on a group of basic sectors thereof. The basic sectors are those whose sales are largely exported to points outside the study area and whose scale of operation may be extrapolated on a national or regional basis. It is postulated that changes in the basic sectors are soon followed by corresponding changes in the rest of the local economy. A historical analysis is made to yield a multiplier ratio between total employment and basic employment. Thus, if the national or regional market's future demands on the basic sectors are estimated exogenously, the future of the local economy can be predicted. The method should not be used for areas whose basic sectors form a small part of the overall economy. The method may be good for an industrial city like Detroit, but is not suitable for places as diversified as New York City.

The most promising approach for predicting long-term economic development is known as Leontief's input-output model. Used in a New York area study[7] and in the Northeast Corridor Transportation Study,[8] this model exhibits explicitly the interactions of all industrial sectors by equating the output of each sector to the inputs of all other sectors. The method is illustrated here by the model used in the New York study.

The local area economy is divided into N industrial sectors. For each sector's product, it is necessary to distinguish between output absorbed by the local area market and output exported to the national market. Let

$x_i(t)$ = output per unit time of ith local industry sector at time t,

$z_{ij}(t)$ = purchases per unit time by jth local industry from ith local industry at time t,

$z_{ic}(t)$ = purchases per unit time by consumers from ith local industry at time t,

$z_{ig}(t)$ = purchases per unit time by government from ith local industry at time t,

$z_{in}(t)$ = purchases per unit time by the national market from ith local industry at time t.

There are N equations of continuity that together express the required balance between the output rates of the N industry sectors and the consumption of their products by local and national markets:

$$x_i(t) = \sum_{j=1}^{N} z_{ij}(t) + z_{ic}(t) + z_{ig}(t) + z_{in}(t) \qquad i = 1, 2, \ldots, N \qquad (1)$$

In the absence of expansion or contraction of productive capacity, the jth sector's purchases from the ith sector are assumed to be proportional to the output of the jth sector. The constant of proportionality a_{ij}, giving raw material i required per unit of product j, can be viewed as constant only as long as there is no significant technological change. If industry sector j expands, purchases from sector i are assumed to be proportional to the rate of change of output in industry j. Again the constant of proportionality c_{ij} remains constant only in a technologically stagnant society. However, a_{ij} and c_{ij} may perhaps be viewed as near constant over a period of several years. If so,

$$z_{ij}(t) = a_{ij}x_j(t) + c_{ij}\frac{dx_j(t)}{dt}. \qquad (2)$$

It may be assumed that a contracting industry is unable to sell its capital assets at significant prices and hence that $c_{ij} = 0$ when $dx_j(t)/dt$ is less than zero.

A replacement of the derivative in Eq. 2 by a difference is rather inevitable because the system of equations being developed is too complicated to admit anything but a numerical solution. Thus

$$\frac{dx_j(t)}{dt} \simeq \frac{x_j(t) - x_j(t - \Delta)}{\Delta}. \qquad (3)$$

Let $P(t)$ = local area population at time t and let $y(t)$ = rate of personal income per unit time in the local area at time t. Consumer purchases are assumed to be proportional to $P(t)$ and to $y(t)$ with time-independent constants of proportionality m_i and f_i:

$$z_{ic}(t) = m_i P(t) + f_i y(t) \tag{4}$$

Government purchases from industry i depend on population and it is assumed that

$$z_{ig}(t) = a_{ig}(t)P(t) + b_{ig}(t). \tag{5}$$

The parameters a_{ig} and b_{ig} are made time-dependent, thus providing a mechanism for modeling the effects of varying government policies on the economy.

If Eqs. 2–5 are substituted into Eq. 1, one obtains

$$x_i(t) = \sum_{j=1}^{N}\left[\left(a_{ij} + \frac{c_{ij}}{\Delta}\right)x_j(t) - \frac{c_{ij}x_j(t-\Delta)}{\Delta}\right]$$

$$+ [m_i + a_{ig}(t)]P(t) + f_i y(t) + b_{ig}(t) + z_{in}(t). \tag{6}$$

Purchases by the national market from local industry, $z_{in}(t)$ for all i, must be estimated by some means outside of the local region model exhibited here. A similar model of the national economy, as a self-contained unit, might be used for this purpose. Extrapolation of past trends is a more likely approach, when a regional study is made, largely because otherwise the study costs can become preposterously large.

The coefficients a_{ij}, c_{ij}, m_i, f_i, a_{ig}, and b_{ig} are viewed as known quantities and must be obtained by observations of past behavior and/or by postulate. Thus at this stage there are N Eqs. 6, with $N + 2$ unknowns, $x_i(t)$ for $i = 1, 2, \ldots, N, y(t)$, and $P(t)$. A further formulation is required to relate income and population to employment, which latter is a function of production. Let

$$E(t) = \text{total area employment at time } t,$$

$e_j(t), e_g(t), e_h(t) = $ regional employment in jth local area industry, in local government, and in private households at time t.

Assume that

$$e_j(t) = h_j(t)x_j(t), \qquad e_g(t) = q(t)P(t) + s(t),$$

and that $e_h(t)$ is determined exogenously. The parameter $h_j(t)$ is the number of employees per unit of product in industry j. $q(t)$ and $s(t)$ are subject to deliberate public policy. Then

$$E(t) = \sum_{j=1}^{N} h_j(t)x_j(t) + q(t)P(t) + s(t) + e_h(t). \tag{7}$$

Let

$$P(t) = g(t)E(t), \qquad y(t) = k(t)E(t),$$

with $g(t)$ and $k(t)$ again determined exogenously.

There now are a set of $(N + 1)$ equations, 6 and 7, with $(N + 1)$ unknowns, $x_i(t)$ and $E(t)$. After all the parameters have been estimated, either by observation or by planning fiat, these equations may be solved successively for moments separated by the time interval Δ. The algebraic matrix is likely to be very large, requiring the use of an electronic computer.

It is obvious that this model, though complex, is a gross simplification of reality. Economic progress is assumed to be orderly and continuous, not affected by panics, wars, or other irregularities. The results can only be worse than the assumptions made; yet, regrettably, the present state of economic forecasting is not encouraging to much greater elaboration.

LAND USE. The several methodologies for land use forecasting have been recently formulated, and there is available little verification or comparison on their accuracies. All require extant land use inventories and independent forecasts of the overall demand for land, the latter expressed in terms of building construction rates for the area as a whole. The overall growth forecasts should be outputs of the population and economy forecasting efforts.

The models may be classified as being either simple extrapolations of past trends or simulations of assumed human behavior, both constrained by the available resources. The simulations may be further classified into deterministic models, wherein people are assumed to be guided by explicit rational economic principles and probabilistic models, wherein people are assumed to be guided by only partially rational centripetal urges. Examples will now be given.

The easiest approach to a simple extrapolation is to develop a multiple linear regression equation of the form

$$Y = c_0 + c_1 X_1 + \cdots + c_n X_n, \tag{8}$$

where Y is the predicted change in a zone's inventory of dwelling units, retail space, factory jobs, or other land use measures; $X_1 - X_n$ are independent predictor variables, such as vacant land area, land value, distance or time to employment opportunities, distance or time to the centroid of the urban area, zoning indices to represent the density and types of construction allowed, and availability of transit. To allow flexibility, the predictor variables might be nonlinear functions of these quantities. The problem is to determine the constants $c_0 - c_n$ on the basis of past trends. Thus land use surveys must be available for at least two moments in time.

For example, in applying this technique to Greensboro, North Carolina, Swerdloff and Stowers[9] developed the following predictor:

$$Y = -2.3 + 0.061 X_1 + 0.00066 X_2 + 1.1 X_3 - 0.11 X_4 - 0.0073 X_5,$$

where Y = logarithm of change in dwelling units, 1948–60, per unit of dwelling land, X_1 = a residential zoning index for 1948, X_2 = percentage of land in residential use in 1948, $X_3 =$ logarithm of time to employment centroid for 1960, $X_4 =$ dwelling unit density in 1948, and $X_5 =$ percentage of land in industrial use in 1948. The coefficient of correlation for the area's several zones was a somewhat disappointing 0.61. If this formula were now to be used to predict housing growth in the equivalent period 1960–1972, it would be necessary to predict 1960 travel times in some external manner. Naturally, the sum of the predicted growths, for all zones, should be normalized to be equal to the overall predicted regional growth.

A somewhat intuitive approach to extrapolation, requiring experienced judgment, is known as the density–saturation gradient method. Used in the Chicago Area Transportation Study,[10] the method consists of an adjustment for consistency of two extrapolated curves. One curve is for dwelling unit density as a function of travel time from a central point. The other is for saturation, the proportion of available residential sites actually occupied by residences, also as a function of travel time from the same central point.

One deterministic simulation is based on the hypothetical rule that each household seeks to maximize its rent-paying ability, or equivalent if the residence is owner-occupied, while land owners seek to maximize the returns on their properties. A linear programming formulation of this maximal principle has been developed, with constraints to ensure that all households are located somewhere and that the total land used in each zone does not exceed the available land area. As used in the Penn–Jersey Transportation Study,[11,12] the model is quite complicated and has been coded for computer analysis.

There are two main probabilistic simulation approaches to land use prediction. The first, known as the accessibility or gravity model, postulates that the likelihood of building on a site in a given zone is largely a function of the zone's distance or travel time separation from places of employment or commerce that give rise to trips originating from the study zone. The second, known as the intervening opportunity model, postulates that the likelihood of building on a site in a given zone is a monotonically decreasing function of the number of equivalent available sites nearer than it to the centroid of all such sites. Since the predictors for both models change with time, it is appropriate to develop time–dependent models.

The accessibility model, first used in a time-independent manner for transportation study by Hansen,[13] has the form

$$g_i = g_t \frac{A_i^a V_i}{\sum_i A_i^a V_i},\tag{9}$$

where $g_i =$ the forecast expected growth rate for zone i, $g_t = \sum g_i =$ total regional growth rate, determined exogenously, $A_i =$ an accessibility index for zone i with empirically determined constant exponent a, and $V_i =$ vacant land area in zone i. The accessibility index has the general form

$$A_i = \sum_j E_j F_{ij},$$

where $E_j =$ a measure of activity in zone j such as total employment, and $F_{ij} =$ "friction" of time separation between zones i and j. F_{ij} may be made proportional to the actual number of trips between zones i and j, as surveyed or predicted exogenously. If nothing is known of the travel patterns between i and j, an approximate measure might be $F_{ij} = 1/T_{ij}^b$, where T_{ij} is the travel time between zones i and j.

Intervening opportunity models have been proposed by Stouffer,[14] Schneider,[15] Sherratt,[16] and Helly.[17] The last of these is a time-dependent formulation. The general approach is as follows. Let there be one centroid, and let each builder's site selection proceed by considering first the sites nearest the centroid. Let $M(r)$ be the number of sites within a radius r about the centroid. Because sites are numerous in a real city, integer-valued $M(r)$ is treated as a continuous variable. Let dM be the number of sites between M and $M + dM$. Let $p\,dM$ be the probability that a given house's site is selected in the range dM, given that it was not selected nearer to the centroid than M. p is originally considered to be a constant parameter. Then the probability of not selecting a site in the range $(0, M + dM)$ is given by

$$Pr(M + dM) = Pr(M)(1 - p\,dM),$$

with the time-independent solution

$$Pr(M) = e^{-pM}. \tag{10}$$

If a total of N_T houses are built, the number built on the M most central sites is $N(M) = N_T[1 - Pr(M)]$, and the density $n(M)$ of houses is

$$n(M) = \frac{dN(M)}{dM} = pN_T e^{-pM}.$$

This time-independent formulation does not ensure that $N(M) \leq M$. This can be overcome by a reformulation in which a site count variable m is introduced, with all available sites at time zero described, in order of decreasing accessibility from the centroid, by the integers $1, 2, \ldots, m, \ldots$. Again the site count is treated as a continuous variable. Now one defines $M(m, t)$ as the number of sites, no less desirable than site m, still vacant at time t. Note that $M(m, 0) = m$, and that $M(m, t)$ has the same meaning as the time-independent M. Thus $Pr(M) = \exp(-pM)$, as in Eq. 10, except that now M is time-dependent.

Let $N(t)$ be the total number of sites occupied at time t, determined exogenously, so $M(m, t)$ is equivalent to $M(m, N)$. In an infinitesimal time dt, dN housing units will be built. Of these, a fraction $[1 - Pr(M)]$ are expected to be built on the M most accessible, still vacant available sites. The number built in dt on these M sites equals the number of these sites that become occupied and thus removed from the vacant site stock. Hence

$$\frac{dM(m, t)}{dt} = \frac{dM(m, N)}{dt} = -\frac{dN}{dt}[1 - Pr(M)].$$

Substituting from Eq. 10, one obtains

$$\frac{dM}{dN} = -(1 - e^{-pM}), \tag{11}$$

with solution

$$M(m, N) = \frac{1}{p}\log\left[(e^{pm} - 1)e^{-pN} + 1\right]. \tag{12}$$

Let

$F(m, N)$ = expected fraction of sites occupied in range $(m, m + dm)$

$$= \frac{d[m - M(m, N)]}{dm}$$

$$= \frac{e^{pN} - 1}{e^{pN} + e^{pm} - 1}. \tag{13}$$

Figure 4 exhibits the form of Eq. 13 for various values of pN and of pm. Since the abscissa is in units of pm, large values on it correspond to relatively inaccessible sites with large m. The exhibited family of curves shows the progression of the city from its beginning with few houses, pN small, to its later stages, pN large, when most of the more desirable sites have been built upon. Because of the early saturation of the more desirable sites, later building occurs almost entirely on the outer fringes. This argument is correct for new site development only, not for the rebuilding of earlier structures.

More elaborate analyses of Eq. 11 are possible, the most important of which includes a time lag between housing decisions and the ultimate construction which then affects later housing decisions.

The accessibility and intervening opportunity models, as presented above, probably are the most promising land use prediction approaches. The accessibility model has the disadvantage of not offering any description, however simplified, of the actual process of site selection. It does have the advantages of (1) providing in F_{ij} a means, more precise than simple ordering, of classifying sites (which may be advantageous when there is a great physical gap between two successive sites as imposed, for example, by a green belt or

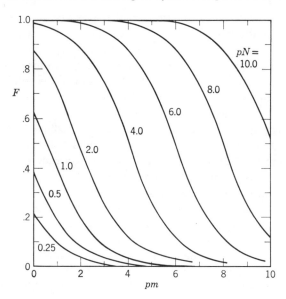

FIGURE 4. Housing site occupancy. F is probability that a site is occupied; m is number of sites in decreasing order of accessibility; N is total number of occupied sites; p is probability that a "considered" site is accepted.

river), and (2) permitting a simple procedure for ordering opportunities in a megalopolis with several centroids. The second advantage may of course be incorporated in the intervening opportunity model by using a general accessibility index as a means for ordering the sites.

The above discussions have been focused on residential land use prediction. Similar models may be used for commercial or industrial development, although it should be kept in mind that these are zoned much more severely by law and custom and that sometimes stringent physical demands are made on location and accessibility to ensure the success of a commercial enterprise. Thus commercial and industrial development may be better predicted, at times, by individual consideration of each major economic activity, its expansion potential, and the land resources explicitly available to it.

Prediction of the Number of Trips Originating from a Zone

Trips on a road network are made by drivers and passengers of private automobiles, by people riding buses and taxis, and by commercial users.

The private trips usually predominate, except perhaps in the central business districts of the most congested cities with major transit facilities, such as New York or Boston. It is reasonable that the number of trips taken by a private household will be a function of the ages, occupations, location relative to possible destinations, income, and automobile ownership of that household. Then, if appropriate land use, population, and economy prediction has been accomplished so as to estimate the number and characteristics of households in each zone, one may deduce the total number of trips arising therein.

The main problem is to predict the number of trips, classified by major purpose, originating from residences. If these are predicted correctly, the remaining private trips, being mostly return journeys to homes, present little difficulty. The most useful predictor variables for residence-based trips have been population density, distance from the central business district, household income, mean household size, and car ownership.

Often population density and household income are sufficient for effective prediction. Most other potential predictors tend to be closely correlated with these. For example, auto ownership was predicted on the basis of these two variables for the New York metropolitan area.[18] A fitted linear regression equation gave a correlation coefficient of $R = 0.95$, when applied to more than a thousand zones in this area. Naturally, for future prediction, one must keep in mind that auto ownership tends to increase with time, though many areas are fairly near to saturation in this regard. Similarly, distance from the central business district is closely correlated with population density, particularly if the distance variable is made travel time. However, the relationship may vary widely between cities because of different development histories and zoning restrictions.

For modeling purposes, the number of trips per household is modeled by a correlation function of predictor variables, fitted to the facts as observed at present, either in the prediction area or in an area as similar to it as possible. Although it is reasonable to expect the form of such a relationship to hold true for some years ahead, past trends ought to be analyzed and extrapolated to deduce whether the overall average number of trips per household is likely to increase with time.

A vast literature on specific area trip generation exists, particularly in the *Highway Research Record*.[19] Unfortunately, most of this is limited to useful correlations in numerical form for a number of individual cities. There has been little generalization or analysis of the predictive value of generation models. The latter is, of course, possible, at least in principle, being best accomplished by predicting the present on the basis of past data.

Trip Distribution

As defined previously, trip distribution is the apportionment of trips, originating in one place, among all possible destinations. The presently used methodologies for trip distribution seek to deduce the relative likelihood of a given destination by use of rather simple functions involving only the amount of activity at the destination and the relative difficulty of traveling to it.

It has been found that the accuracy of a trip distribution model is much enhanced if the model is applied separately to each of several trip categories, such as work (i.e., commuting), shopping, business, or school. The extent to which such a breakdown is made depends on the resources available for obtaining and processing the pertinent data.

Any trip distribution model must be checked and calibrated by application to present or past actual behavior. The effort required to obtain present travel patterns, by observation or questionnaires, usually represents a most dominant cost of any network travel prediction effort.

In the models to be described, it is assumed that the area of interest is divided into N zones and that these N zones encompass the entire universe of interest. Hence trips to and from areas outside the study area must be handled by formally labeling the outside world as one or more of the N zones.

The following notation will be used:

t_{ij} = number of trips, originating in zone i and terminating in zone j, during a specified time period such as an average weekday or rush hour, as predicted by a trip distribution model for an observable (i.e., present or past) base period.

t_{Eij} = an estimate of t_{ij} obtained, exogenously to the trip distribution model, by survey, observation, or synthesis.

$t_i = \sum_{j=1}^{N} t_{ij}$ = number of trips originating in zone i during the specified time period, as predicted by a model for an observable base period.

$t_{Ei} = \sum_{j=1}^{N} t_{Eij}$ = an estimate of t_i obtained exogenously to the trip distribution model.

T_{ij} = number of trips, originating in zone i and terminating in zone j, during the specified time interval, as predicted by the trip distribution model for a future prediction period.

$T_i = \sum_{j=1}^{N} T_{ij}.$

T_{Ei} = an estimate of T_i obtained, exogenously to the trip distribution model, by application of a trip generation analysis.

If the period of interest is one day (24 hr), then it is generally assumed that $t_{ij} = t_{ji}$, $t_{Eij} = t_{Eji}$, and $T_{ij} = T_{ji}$. A complete tabulation of any of these quantities, for all i and j, is generally referred to as a trip table and may be displayed conveniently in matrix form.

Growth Factor Models

Suppose

$$G_i = \frac{T_{Ei}}{t_{Ei}} = \frac{\text{predicted future trips originating in zone } i}{\text{observed present trips originating in zone } i}.$$

If one attempts to distribute trips among zones entirely on the basis of growth factors G_i ($i = 1, 2, \ldots, N$), then about the simplest meaningful approach would be the average growth factor method:

$$T_{ij} = t_{Eij}\left(\frac{G_i + G_j}{2}\right). \tag{14}$$

Such a simple extrapolation, ignoring any change induced by changes in the transportation network on travel times, is likely to be useful only for short-term prediction.

To use this method, one must be prepared to iterate to obtain internally consistent results. From Eq. 14, one obtains $T_i = \sum_{i=1}^{N} T_{ij}$, which is not likely to be equal to T_{Ei}. Yet T_{Ei} is the basic estimate of travel volume for zone i. To adjust the T_{ij} terms so T_i approaches T_{Ei}, one may view the value obtained in Eq. 14 as $T_{ij}^{(1)}$ in an iterative procedure where

$$T_{ij}^{(n+1)} = T_{ij}^{(n)}\left(\frac{G_i^{(n+1)} + G_j^{(n+1)}}{2}\right)$$

and

$$G_i^{(n+1)} = \frac{T_{Ei}}{\sum_{j=1}^{N} T_{ij}^{(n)}} \qquad \text{for} \quad n \geq 1.$$

The iteration is repeated until some step m where

$$(T_{Ei} - T_i^{(m)}) = \left(T_{Ei} - \sum_{j=1}^{N} T_{ij}^{(m)}\right)$$

is deemed to be sufficiently close to zero.

A somewhat more complex growth factor formulation is the nonlinear Fratar model[20] in which

$$T_{ij} = t_{Eij} G_i G_j \left(\frac{t_{Ei}}{\sum_{n=1}^{N} t_{Ein} G_n}\right). \tag{15}$$

Since $G_i = (T_{Ei}/t_{Ei})$, the Fratar model ensures that $T_i = T_{Ei}$, and no iteration is required to achieve this. However, unlike the average growth factor approach, there is no forcing of $T_{ij} = T_{ji}$, so iteration may be required if these differ substantially.

There is some evidence[21] to suggest that the Fratar method offers no improvement over the simpler average growth factor approach.

Gravity Model

The growth factor models do not explicitly take into account the relative travel costs, in distance, time, or money, of alternative destinations. Thus a growth factor approach is totally useless for assessing the changes in travel patterns resulting from major changes in the transportation network occasioned by new construction of facilities or by new schedules for public transit.

The gravity model[22] is a first-order approach wherein it is assumed that the number of trips between two zones is proportional to the total number of trips originating in the origin zone i, to the number of potential trip ends in the terminating zone j, and to a cost function F_{ij}, often called the "friction" between the two zones. If the zones are near to each other, F_{ij} is large; if not, F_{ij} is small. For observed base period travel patterns, the model takes the form

$$t_{ij} = t_{Ei}\left(\frac{A_j F_{ij}}{\sum_{n=1}^{N} A_n F_{in}}\right), \tag{16}$$

where A_j is a measure of the number of attractions or useful trip ends in zone j. The sum in the denominator is a normalizing factor that ensures that

$$\sum_{j=1}^{N} t_{ij} = t_{Ei}.$$

If some specific class of trips is to be analyzed, A_j may be defined appropriately to that class. For example, for work trips, A_j might be the number of jobs in zone j. If all types of trips are lumped together, and if it is assumed that $t_{ij} = t_{ji}$, then $A_j = t_{Ej}$ might be appropriate.

The main problem in making Eq. 16 operational is to find a function F_{ij} that can be applied to all i and j. The practical approach is to assume, at least initially, that $F_{ij} = F(d_{ij})$, where d_{ij} is the travel time from zone i to zone j. The model is calibrated by developing an empirical function $F(d)$, which, when used in Eq. 16, will minimize the sum of the squared discrepancies

$$(t_{ij} - t_{Eij})^2$$

between the model and the real world. If such a function is developed and found to work reasonably well for present, observed travel patterns, then it is assumed that this empirical $F(d)$ will remain appropriate for future prediction. If so, future trip distribution is calculated from

$$T_{ij} = T_{Ei}\left(\frac{A_j F_{ij}}{\sum_{n=1}^{N} A_n F(d_{in})}\right),$$

where the form of $F(d)$ is the same as for the present and where A_j, T_{Ei}, and d_{ij} are the best available estimates of future attractions, trip generation, and travel times.

When the study area has special localized features, such as a major barrier (i.e., a river) or a costly toll facility, F_{ij} may usefully be expanded to

$$F_{ij} = F(d_{ij})R_{ij},$$

where R_{ij} is a correction factor to account for the special impediments. $R_{ij} = 1$ if the trip between i and j does not impinge on such a troublespot.

For very rough calculations and for conceptual purposes, one may try

$$F(d_{ij}) = \frac{1}{d_{ij}^c},$$

where c is a constant to be fitted by use of presently available trip distribution data. The term "gravity model" arises from the parallel to gravitation for the special case of $c = 2$.

Intervening Opportunity Model

The intervening opportunity model[23] is based on the supposition that, for each trip, all appropriate possible destinations are ordered on the basis of travel time from the trip origin, with the nearest one considered first, the second nearest second, and so forth. One assumes that there is a constant probability p that a considered site is accepted as the trip end.

This model is an exact analogy to the time-independent intervening opportunity model for land use prediction.

Let $M(d)$ be the number of suitable destinations nearer than d to the trip origin. Usually d is viewed as the travel time. Because the number of possible destinations usually is large, $M(d)$ is treated as a continuous variable. Then $P(M)$, the probability of refusing the nearest M sites to the origin, can be deduced from

$$P(M + dM) = P(M)(1 - p\,dM),$$

with solution

$$P(M) = e^{-pM}.$$

On this basis,

$$t_{ij} = t_{Ei}\left[\left(\begin{array}{c}\text{Probability of not} \\ \text{accepting any destination} \\ \text{nearer than zone } j\end{array}\right) - \left(\begin{array}{c}\text{Probability of not} \\ \text{accepting any destination} \\ \text{in zone } j \text{ or nearer} \\ \text{than zone } j\end{array}\right)\right]$$

so

$$t_{ij} = t_{Ei}\{\exp(-pM_j) - \exp[-p(M_j + m_j)]\}, \tag{17}$$

where M_j is the number of destinations nearer to the origin than the destinations in zone j, and m_j is the number of destinations in zone j. Equation 17 is used to calibrate the model to present data by finding the best possible value of p. Once this is done, the same equation is used to estimate T_{ij} in terms of T_{Ei}.

This model is not altogether consistent in its overall results for a study area. For a discussion of some of the problems, see Ref. 24.

Minimum Entropy Formulation

The gravity model on page 263 is a proportional model in the sense that the number of trips from zone i to zone j is proportional to the total number of trips t_{Ei} from zone i and to the number of attractions A_j in zone j. One may simplify the model to show only these basics by ignoring friction and setting $F_{ij} = 1$ for all i and j. Then Eq. 16 becomes

$$t_{ij} = \frac{t_{Ei} A_j}{\sum_{n=1}^{N} A_n} . \tag{18}$$

On a 24-hr basis, one may assume that $t_{ij} = t_{ji}$ and that $A_j = t_{Ej} =$ the number of trips observed or estimated to originate from zone j. Then the conditions satisfied by the now simplified model are

$$\sum_{j=1}^{N} t_{ij} = t_{Ei},$$

$$\sum_{i=1}^{N} t_{ij} = A_j = t_{Ej}, \tag{19}$$

$$\sum_{i=1}^{N} \sum_{j=1}^{N} t_{ij} = \sum_{j=1}^{N} t_{Ej} = T,$$

where T is the total number of trips in the network.

These conditions formally express the conservation laws that every trip has both a start and a finish and that the total number of trips in the area is equal to the sum of all the zonal trips.

It will now be shown that the same Eq. 18 can be obtained by minimizing the function

$$S = \sum_{i=1}^{N} \sum_{j=1}^{N} t_{ij} \log t_{ij}, \tag{20}$$

subject to the basic conditions of Eq. 19. Since the function S looks like an entropy function, as found in statistical thermodynamics, and is used like an entropy function, this methodology has been called the entropy model.[25]

To minimize Eq. 20 subject to Eq. 19, the method of Lagrange multipliers is used. The required result is obtained by minimizing the composite function

$$F = S - \sum_{i=1}^{N} a_i\left(t_{Ei} - \sum_{j=1}^{N} t_{ij}\right) - \sum_{j=1}^{N} b_j\left(t_{Ej} - \sum_{i=1}^{N} t_{ij}\right) - c\left(T - \sum_{i=1}^{N} \sum_{j=1}^{N} t_{ij}\right),$$

where a_i, b_j, and c are undetermined multipliers. To minimize F, one takes its derivatives with respect to t_{ij}, a_i, b_j, and c and then sets each of the derivatives equal to zero. This process yields

$$\frac{\partial F}{\partial t_{ij}} = \log t_{ij} + 1 - a_i - b_j - c = 0 \qquad (21)$$

and the constraints imposed by Eq. 19. The solution to Eq. 21 is

$$t_{ij} = \exp(a_i + b_j + c - 1). \qquad (22)$$

Substitution of Eq. 22 into Eq. 19, followed by multiplication of the first two relations of Eq. 19 and division of the result by the third relation yields

$$t_{ij} = \exp(a_i + b_j + c - 1) = \frac{t_{Ei}t_{Ej}}{T} = \frac{t_{Ei}A_j}{\sum_{j=1}^{N} A_j}.$$

Since this is the same as Eq. 18, the entropy formulation is shown to be equivalent to the simplified gravity model. It is, of course, possible to postulate more complicated entropy functions to take into account the relative travel times between zones. Although the entropy approach has not been used directly for trip distribution studies, it does present potential for more sophisticated conceptualization and analysis than does the gravity model, so there is reason to explore it further.

A somewhat different formulation and a more extended treatment of entropy distribution models is given by Wilson.[26] The entropy viewpoint is criticized severely by Beckmann and Golob[27] on the ground that identical results can be obtained by direct utility arguments, uncomplicated by resort to the metaphysical apparatus of the second law of thermodynamics.

Modal Split and Diversion Assignment

Modal Split

When travelers have available both a private car road network and a public transit system, proper assignment of traffic requires estimates of the *modal splits* of travelers between these alternatives. Whether a given person chooses one or the other is likely to depend on a fairly complicated individual evaluation of relative costs, travel times, comfort, convenience, reliability, status value, and the like. Naturally it also depends on whether the individual has a car and on his financial resources. Many of these factors are influenced dramatically by the spread and level of service provided by public transit; in the absence of transit there is no choice.

Modal split models usually predict the proportion of travelers diverted to public transit as a function of suitable predictor variables, such as relative cost and travel time. Such models can be applied in a gross way to predict the modal split for the group of all travelers from a zone before any distribution is attempted for the various destinations of these trips. A more accurate, and inevitably more cumbersome, approach is to first distribute trips among possible destinations and then to make a modal split estimate on the basis of both specific origin and destination so that the specific availability of public transit may be considered.

One may be prepared to postulate a relationship $F = F(X_1, X_2, \ldots, X_k)$, where F is the fraction of travelers diverted to transit and the X's are predictor variables, such as differences or ratios of travel times or costs between the modes. If so, the problem becomes one of calibrating the parameters in F so that this function most closely reproduces observed behavior. This calibration is known as discriminant analysis and is described in standard texts, for example, Ref. 28.

The procedure is particularly simple if the discriminant function F is taken to be linear. Then one seeks coefficients a_i for a best least-squares fit of data to the function

$$F = a_0 + a_1 X_1 + \cdots + a_k X_k.$$

It should be kept in mind that the meaningful range of F is between 0 and 1.

To find the coefficients a_i, one proceeds as follows.[29] For observations j, $j = 1, \ldots, N$, let x_{ij} be the observed predictor variable and let $Y_j = 1$ if transit was chosen. Let $Y_j = 0$ if a private automobile was chosen. Let

$$G = \sum_{j=1}^{N} \left(a_0 + \sum_{i=1}^{k} a_i X_{ij} - Y_j \right)^2.$$

The values of a_i giving the best least-squares fit are found by solving the set of k linear equations

$$\frac{\partial G}{\partial a_i} = 0 \qquad \text{for} \quad i = 1, 2, \ldots, k.$$

Reference 30 exhibits one practical study done in this manner.

If one desires to model the travelers' decision processes more directly, one may postulate that each traveler evaluates a generalized cost, or "disutility," for each travel mode available to himself. He then travels on the mode he decides is cheapest. Because travelers do not all come to the same conclusions regarding costs, some proportion of them will take the mode(s) which the majority deems to be more expensive.

Let $c_1, c_2, \ldots, c_k$ be disutility predictor variables such as travel time, time spent walking, or money cost. The disutility of a trip, between two points via a given mode, may be assumed to have the linear form

$$C = \sum_{i=1}^{k} a_i c_i,$$

where the a_i's are coefficients to be determined. Then the difference in cost between two modes I and II is

$$\Delta C = \sum_{i=1}^{k} a_i (c_i^{\mathrm{I}} - c_i^{\mathrm{II}}).$$

One argues that each traveler evaluates ΔC and chooses mode I if ΔC is less than zero, or mode II if ΔC is greater than zero. Among any group of travelers, the individually estimated c_i's will vary. Therefore they are viewed as random variables. Calibration of the model consists of finding suitable distributions for the c_i's and suitable values for the a_i's so that the model will best reproduce observed behavior. To simplify matters, it is reasonable and standard to assume that the distributions of the c_i's are approximately normal (Gaussian). This approach is discussed by Warner,[31] and used by Stopher,[32] Bevis,[33] and Pratt.[34]

Mitchell and Clark[35] offer a somewhat different procedure for cases where there are insufficient data to quantify equivalencies between different types of disutilities. The approach is to extrapolate from observed situations under the covering assumption that "for journeys on which any two modes carry the same number of travelers the generalized cost for the two modes is the same." It is necessary to find several cases of such an equality, for different travel distances, before one can extrapolate the relative time costs that travelers ascribe to each mode.

Diversion Assignment

A special case of the assignment problem, to be considered generally in the next subsection, occurs when one facility is to be newly built or changed dramatically. A *diversion assignment model* is any technique for predicting how much traffic will be diverted to the new facility. The usual approach, formally the same as for modal splits, is the use of a statistical discriminant function for the proportion of traffic diverted to the new facility as a function of predictor variables such as travel times and costs via both alternatives. The ratio of travel times between the alternatives has been used as the sole predictor in some simple models as, for example, the one in the Bureau of Public Roads computer program package for traffic assignment.[36]

Assignment

Practical experience in the assignment of traffic onto the graph of a network has been limited mostly to an all-or-nothing procedure, wherein all traffic between two zones is assigned to the same route. We shall exhibit this viewpoint, the methodology applied therefor, and discuss some of the efforts toward multiple route assignment. In addition, this section presents a land use accounting approach to overall planning wherein assignment is done on a general location basis rather than onto individual roads.

All-or-Nothing Viewpoint

A widely held viewpoint is that the vast majority of travelers choose the shortest time routes between their origins and destinations. There is no evidence to cast doubt on this claim except, possibly, cases of recreational travel. The simplest way to implement the shortest route philosophy in a model is to assign all traffic from one zone to another to one route. Though simple to analyze, such an all-or-nothing model has faults that contribute toward substantial inaccuracies in the predictions made when the model is used.

The main difficulty arises when two or more routes between two points offer virtually the same travel times. This situation arises when the travel time is short compared with the terminal times associated with obtaining and parking a vehicle. Further, when the environment is congested, traffic spreads over all available routes. Those routes which are the fastest in the absence of traffic attract such a fraction of the overall volume as to result in roughly equal travel times on them as on the competing routes which would be slower in the absence of heavy traffic.

Other causes for poor accuracy are not inherent in the method. They arise from such simplifications as the use of zones rather than individual destinations and the averaging of substantial time intervals in place of a time-dependent analysis. When the centroid of a zone is viewed as the origin for all trips starting in that zone, errors are inevitable and may be substantial, particularly in the assignment of trips to the immediately adjacent zones. Furthermore, the selection of a route in a congested area may be critically dependent on the particular road access of the terminal in which a vehicle is parked. A traffic assignment which averages behavior over a substantial time period will not account for the possibility that the shortest time route may be different during a rush hour than at other times.

In some cases minimum time is not the criterion adopted by drivers. This will be the case if there are tolls on some but not all competing routes, or if travel distances vary very much among the routes. However, these complications, if recognized and quantified, are easy to represent in a model by making the cost of a link a function of money or distance as well as of time.

Finally one should point out that the value of time may be quite different to different groups of travelers. Hoshino[37] points out that inclusion of this consideration can lead to results significantly changed from those obtained by the simpler aggregation argument of the following pages.

All-or-Nothing Assignment without Capacity Restraint

The following discussion uses the network graph terminology introduced on page 244 of Section I.

Capacity restraint is the term used to describe the use of a variable travel time for the cost of traversing a network link. The cost would rise with an increase in the traffic assigned to that link. If one simplifies by selecting one constant travel time cost for each link of the network, independent of the traffic volume assigned, then the absence of capacity restraint makes possible a very simple and rapid assignment procedure.

Suppose that one is given

1. A network model in the form of a link–node graph. Each network zone has associated with itself one node of the graph as the origin of all traffic originating from it. It also has one node as the destination of all trips terminating within the zone. Possibly one node may serve both functions.
2. A cost c_{ij} specified for each link (i, j) of the network. Here i and j are, respectively, the origin and the destination nodes of the link.
3. A trip table obtained by survey or by a trip distribution model. If the number of trips from zone i to zone j is t_{ij}, then the trip table is a matrix giving t_{ij} for all nodes i and j that are associated with trip origins or destinations. (Note that a sensible labeling procedure for nodes is to assign the numbers from 1 to N to the N zonal origin nodes, the numbers $N + 1$ to $2N$ to the N zonal destination nodes, and $2N + 1$ on up to the remaining network nodes.)

An all-or-nothing assignment of the trip table traffic consists of finding the minimum cost route, or chain of links, from zone origin i to zone destination j, and then adding the number of trips t_{ij} to the traffic on each link of this chain. The procedure is repeated for every zone pair. The only nontrivial problem is to find the minimum cost chains.

A very simple algorithm for developing minimum cost chains, reasonably suited to efficient computer programming, is given in Ref. 38. This method, now to be described, develops a tree which exhibits the minimum cost chains from a given starting node to all the nodes of the network. The method requires that all times or costs c_{ij} are greater than or equal to zero. The method can be visualized most readily by consideration of an example. Consider the network graph of Figure 5, which consists of 10 nodes and 14

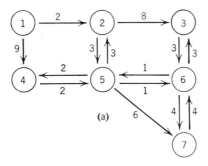

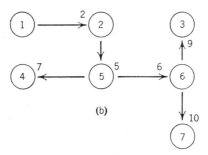

FIGURE 5. A minimum time network
tree. (a) The network. (b) Tree from
node 1.

directed links. Each link has a travel time cost shown next to it. We now
construct a minimum cost tree from node 1 to all nodes of the network.

First, one prepares a table which lists, for each node, all the links
originating from that node arranged in order of increasing cost:

0
1   2   3   4   5   6   7

$(1, 2) - 2$ $(2, 5) - 3$ $(3, 6) - 3$ $(4, 5) - 2$ $(5, 6) - 1$ $(6, 5) - 1$ $(7, 6) - 4$
$(1, 4) - 9$ $(2, 3) - 8$ $(5, 4) - 2$ $(6, 3) - 3$
                                                      $(5, 2) - 3$ $(6, 7) - 4$

STEP 1.   Find the nearest node to the origin.

Look at the origin column 1. Since all links from node 1 are listed, the
shortest of these, (1, 2), and node 2 at its end must be in the minimum time

tree. To document this conclusion, circle the link $(1, 2)$. To prepare for the next step, write the time to node 2 above column 2 and cross out links terminating at node 2 wherever they appear in the table. Since the minimum time tree to node 2 is $(1, 2)$, all other links to node 2 are not in the minimum time tree. The table now looks like this:

0	2					
1	2	3	4	5	6	7

$(\!(1, 2) - 2\!)$ $(2, 5) - 3$ $(3, 6) - 3$ $(4, 5) - 2$ $(5, 6) - 1$ $(6, 5) - 1$ $(7, 6) - 4$

$(1, 4) - 9$ $(2, 3) - 8$ $\qquad\qquad\qquad\qquad$ $(5, 4) - 2$ $(6, 3) - 3$

$\qquad\qquad\qquad\qquad\qquad\qquad\qquad$ $\cancel{(5, 2) - 3}$ $(6, 7) - 4$

$\qquad\qquad\qquad\qquad\qquad\qquad\qquad$ $(5, 7) - 6$

STEP 2.   Find the second nearest node to the origin.

There are at most two candidates for the second nearest node: the nodes, not yet on the tree, nearest to the two nodes already on the tree. Thus one must examine the first entries, not already circled or crossed out, in the columns for nodes 1 and 2. The two nodes are 4, a time 9 from the origin, and 5, a time $2 + 3 = 5$ from the origin. Thus, node 5 is the second nearest node to the origin, so it and the link $(2, 5)$ form part of the tree. Circle this link in the table, enter the time from the origin to node 5 above column 5, and cross out all other links to node 5:

0	2			5		
1	2	3	4	5	6	7

$(\!(1, 2) - 2\!)$ $(\!(2, 5) - 3\!)$ $(3, 6) - 3$ $\cancel{(4, 5) - 2}$ $(5, 6) - 1$ $\cancel{(6, 5) - 1}$ $(7, 6) - 4$

$(1, 4) - 9$ $(2, 3) - 8$ $\qquad\qquad\qquad\qquad$ $(5, 4) - 2$ $(6, 3) - 3$

$\qquad\qquad\qquad\qquad\qquad\qquad\qquad$ $\cancel{(5, 2) - 3}$ $(6, 7) - 4$

$\qquad\qquad\qquad\qquad\qquad\qquad\qquad$ $(5, 7) - 6$

STEP 3.   Find the third nearest node to the origin.

There are at most three candidates for the third nearest node, namely, those connected by the shortest links to the nodes already in the tree. These are node 4, with a time 9 to the origin, node 3 with a time $2 + 8 = 10$, and node 6 with a time $5 + 1 = 6$. Thus node 6 is the third nearest node to the origin. Circle link $(5, 6)$, cross out all other links to node 6, and enter the time to node 6 in column 6:

0	2			5	6	
1	2	3	4	5	6	7

$\boxed{(1,2)-2}\,\boxed{(2,5)-3}\,\cancel{(3,6)-3}\,\cancel{(4,5)-2}\,\boxed{(5,6)-1}\,\cancel{(6,5)-1}\,\cancel{(7,6)-4}$

$(1,4)-9\quad(2,3)-8 \qquad\qquad\qquad (5,4)-2\quad(6,3)-3$

$\qquad\qquad\qquad\qquad\qquad\qquad\quad \cancel{(5,2)-3}\quad(6,7)-4$

$\qquad\qquad\qquad\qquad\qquad\qquad\quad (5,7)-6$

STEP 4.   Find the fourth nearest node to the origin.

The candidates are the terminal nodes of the shortest links in the columns of nodes already in the minimum time tree, namely, those connected by link $(1, 4)$ with time 9, by link $(2, 3)$ with time $2 + 8 = 10$, by link $(5, 4)$ with time $5 + 2 = 7$, and by link $(6, 3)$ with time $6 + 3 = 9$. The shortest time is 7, and so we add link $(5, 4)$ and node 4 to the tree. Modify the table as before:

0	2		7	5	6	
1	2	3	4	5	6	7

$\boxed{(1,2)-2}\,\boxed{(2,5)-3}\,\cancel{(3,6)-3}\,\cancel{(4,5)-2}\,\boxed{(5,6)-1}\,\cancel{(6,5)-1}\,\cancel{(7,6)-4}$

$\cancel{(1,4)-9}\quad(2,3)-8 \qquad\qquad\qquad \boxed{(5,4)-2}\quad(6,3)-3$

$\qquad\qquad\qquad\qquad\qquad\qquad\quad \cancel{(5,2)-3}\quad(6,7)-4$

$\qquad\qquad\qquad\qquad\qquad\qquad\quad (5,7)-6$

STEP 5.   Find the fifth nearest node to the origin.

The candidates are the nodes connected to the tree by link $(2, 3)$ with a time $2 + 8 = 10$, by link $(5, 7)$ with a time $5 + 6 = 11$, and by link $(6, 3)$ with a time $6 + 3 = 9$. The shortest time is 9 and so we add link $(6, 3)$ and node 3 to the tree:

0	2	9	7	5	6	
1	2	3	4	5	6	7

$\boxed{(1,2)-2}\,\boxed{(2,5)-3}\,\cancel{(3,6)-3}\,\cancel{(4,5)-2}\,\boxed{(5,6)-1}\,\cancel{(6,5)-1}\,\cancel{(7,6)-4}$

$\cancel{(1,4)-9}\quad\cancel{(2,3)-8} \qquad\qquad\qquad \boxed{(5,4)-2}\,\boxed{(6,3)-3}$

$\qquad\qquad\qquad\qquad\qquad\qquad\quad \cancel{(5,2)-3}\quad(6,7)-4$

$\qquad\qquad\qquad\qquad\qquad\qquad\quad (5,7)-6$

STEP 6.   Find the sixth nearest node to the origin.

The candidates are connected to the tree by links $(5, 7)$, with time $5 + 6 = 11$, and $(6, 7)$ with time $6 + 4 = 10$. The shortest time is 10 and so link

(6, 7) with node 7 are added to the tree:

0	2	9	7	5	6	10
1	2	3	4	5	6	7

$\boxed{(1, 2) - 2}$ $\boxed{(2, 5) - 3}$ ~~(3, 6) — 3~~ ~~(4, 5) — 2~~ $\boxed{(5, 6) - 1}$ ~~(6, 5) — 1~~ ~~(7, 6) — 4~~

~~(1, 4) — 9~~ ~~(2, 3) — 8~~ $\boxed{(5, 4) - 2}$ $\boxed{(6, 3) - 3}$

~~(5, 2) — 3~~ $\boxed{(6, 7) - 4}$

~~(5, 7) — 6~~

The tree is now complete and consists of the circled links and the nodes they interconnect. The completed minimum time tree is shown in Figure 5. Minimum travel time, as obtained along the tree, is shown next to each node.

There are other methods for finding the minimum cost trees, some of which may be more suitable for programming a particular computer or for solving a particular problem. Best known are the Moore algorithm[39] and a linear programming formulation.[40]

When it is desired to study the effects of varying the cost of one link, it is very wasteful of computer time to repeatedly build up a minimum cost assignment from scratch for each cost assumption. Loubal[41] offers a technique for avoiding this. The essence of the method is to construct two minimum cost chains between each pair of zones. One chain is based on an infinite cost for the study link, so that it is not used at all. The other chain is constructed in three parts:

1. The first part is a minimum cost chain from the origin node to the nearer node (in cost) of the node pair connected by the study link.
2. The second part is the study link.
3. The third part is a minimum cost chain from the farther node of the node pair connected by the study link to the destination node.

For any given study link cost, the cost of a chain through it is the sum of the three costs of the three partial chains just described. If this is less than the cost of the other route, which does not pass through the study link, the traffic is assigned to the study link. Thus, once the two alternative routes are determined for each zone pair, no more minimum cost route analysis is required for changes in the study link cost.

All-or-Nothing Assignment with Capacity Restraint

Capacity restraint is said to apply if the travel time or cost of a link is made dependent on the volume of traffic on that link. In general, once this is done, the assignment must be made in an iterative manner. The simplest way to do this is to make successive assignments with fixed link costs, as in the previous subsection, and to recompute link travel times between iterations. To get rapid convergence without oscillations in the solution, some ingenuity may be required in the choosing of new link travel times for iteration n. These ought to be set at some appropriate point between the times used for iteration $(n - 1)$ and those implied by the assignment of iteration $(n - 1)$.

Snell[42] gives an interesting variational technique for an iterative assignment with nonlinear travel time functions. Further theoretical development by Hoshino[37] is of particular interest. The Hoshino model includes the possibilities of multiple route assignment and of the hourly fluctuations within the day.

Multiple Route Assignment

There are two causes for the use of more than one route between origin and destination. One is that, at any given moment in time, two competing routes may have essentially the same travel time or cost. The other is that, over an extended study time interval, some cars choose their routes when traffic is light and others choose their routes when it is heavy. Neither of these cases has been modeled in any direct sense by multiple route assignment procedures developed so far (1969).

For small networks, it might be possible to do detailed time-dependent simulation and achieve realistic splits among routes. Little work has been done on this. For large networks, models have been constructed with the modest aim of somehow obtaining a variety of plausible routes by techniques reasonably susceptible to rapid computer analysis. Arguments as to how these models actually correspond to real traffic tend to be *a posteriori* and are not very convincing. Thus their validity is established not logically but only to the extent that they improve on the accuracy of all-or-nothing assignment.

According to Humphrey,[43] even the quite haphazard spreading of routes among alternatives improves the accuracy of assignment. When checkable iterative all-or-nothing assignments of present-day traffic were made to achieve capacity restraint in ten American cities, the weighted average assignment errors ranged from 33 to 61% after the first iteration, and from 27 to 67% after the third iteration. However, an average of the three iterations yielded an error range of from 27 to 50%, with improved results for eight of the ten cities. Since this averaging has the effect of splitting some of the traffic over up to three routes in a rather haphazard manner, one deduces that any such splitting onto comparable time routes is of value.

Most multiroute assignment models use an incremental approach. For example, the Chicago Area Transportation Study model selects zones, one at a time, at random. An all-or-nothing assignment is made for all traffic from the first selected zone on the basis of travel times in an empty network.

Travel times are then recalculated on the basis of the traffic volumes imposed by the first zone traffic assignment. The process is repeated, one zone at a time, with travel times modified after the addition of each zone's traffic. The argument is that this procedure is somewhat analogous to some assignment at light traffic hours and some at heavy traffic hours.[44]

A very flexible incremental assignment procedure has been proposed and programmed by Martin.[45] Here one assignment increment consists of only some of the traffic from one origin zone to one destination zone, with the amount of traffic assigned at any one iteration available as an input parameter.

Smeed's Land Use Accounting Model

Smeed[46] has developed an assignment model of the simplest possible character, suitable for visualizing the road requirements of alternative modal splits, alternative land use doctrines, and alternative patterns for arterial roads. Instead of assigning traffic onto specific roads, one calculates the area of land required at all locations as a function of trip distribution, travel mode, and amount of travel done in the peak hours.

The original Smeed model was developed to deduce the minimum central business district area of a city as a function of how people travel to work. The minimum ground area required by a person to make a rush hour trip is a function of the mode of conveyance and the distance traveled. The person also requires space to work and to park his car, if this is his mode of travel.

To deduce the ground area required by the traveler, one requires the flow capacity c, in persons per unit time, on a roadway of width w. Then if the transport system is used at peak capacity for a daily time T, the number of persons who can be carried during the peak period on a unit width of roadway is cT/w. The width required by one person is the inverse of this, namely $w/cT \equiv \lambda$. The area of roadway required for one peak period journey of length L is

$$A = L\left(\frac{w}{cT}\right) = L\lambda. \qquad (23)$$

Table 1 gives typical values of λ for various modes of transport on the basis of British experience and on the assumption that the busy period T is of a 2-hr duration. Also shown are the corresponding values of area required for a 1-mile journey.

Suppose that the smallest central business district sufficient to accommodate N workers is represented by a circle of radius r. Suppose further that all work places are uniformly distributed throughout this district and that all places of residence are outside, with circular symmetry. Assume that "in his journey to work, the worker travels along the straight line from his residence to the town center until he reaches radius r. He then travels directly towards the

TABLE 1

CARRYING CAPACITIES OF VARYING TYPES OF ROADWAY AND ASSOCIATED
AREAS REQUIRED PER PERSON DURING PEAK PERIOD FOR A 1-MILE
JOURNEY (DAILY PEAK BUSY PERIOD $= T = 2$ HR)

Mode of Transport	Flow, persons/ft-width/hr	Speed, miles/hr	λ, Feet	Area Required per Person, ft²
Automobiles on urban street	67	15	0.0074	39
44 ft wide with 1.5 passengers per vehicle	95	10	0.00526	28
Automobiles on urban expressway with 1.5 passengers per vehicle	187	40	0.0027	14
Pedestrian way	800	2.5	0.0006	3
Urban railway line	2900	18	0.00017	0.9
	2200	30	0.00023	1.2

Source. Ref. 45.

place at which he works, the position of which is not correlated with the place
at which he lives or with the point at which he reaches the outer part of the
town center."

Smeed deduces that the average of distances from one point at radius r
to all points within the circle is about $1.36r$, if reasonable provision is made
for restrictions imposed on travel by a rectangular street grid system. Hence
the distance traveled in the central business district by N workers is $1.36Nr$
and the area required for roadway is $1.36Nr\lambda$.

Each worker requires a ground space G for work and an area P for parking.
P is assumed to be zero unless the worker travels by automobile. The sum
of the areas required for roadway, for work, and for parking is equal to the
minimum area of the central business district. Hence

$$1.36Nr\lambda + N(G + P) = \pi r^2,$$

with solution

$$r \cong 0.216N\lambda\left[\left(1 + \frac{G + P}{0.147N\lambda^2}\right)^{1/2} + 1\right]. \tag{24}$$

To examine the general behavior of this equation, Smeed set $G = 100$ ft²
as the net ground space required per person for work, $P = 133$ ft² per person
for ground level parking, and $P = 13.3$ ft² per person for multilevel parking.

TABLE 2

RADIUS r OF CENTRAL BUSINESS DISTRICT AND PERCENTAGE OF GROUND
AREA DEVOTED TO (A) ROADWAY, (B) PARKING, AND (C) WORKING
SPACE (IN EACH GROUP OF RESULTS, r IN MILES IS SHOWN ABOVE THE
PERCENTAGES A, B, AND C)

Mode of Travel to Center	Nature of Parking	Area (P) for Parking per Person in Square Feet	Working Population in Town Center		
			50,000	500,000	5,000,000
Urban railway	—	0	0.24	0.76	2.44
$\lambda = 0.00023$			0.3, 0, 99.7	1, 0, 99	4, 0.96
Car on					
expressway	Multilevel	13.3	0.26	0.86	3.16
$\lambda = 0.0027$			4, 11, 85	13, 10, 77	34, 8, 58
Car on	Ground				
expressway	level	133	0.37	1.21	4.24
$\lambda = 0.0027$			3, 55, 42	9, 52, 39	25, 42, 23
Car on					
44-ft street	Multilevel	13.3	0.27	0.97	4.48
$\lambda = 0.0074$			11, 10, 79	31, 8, 61	67, 4, 29
Car on	Ground				
44-ft street	level	133	0.38	1.32	5.47
$\lambda = 0.0074$			8, 53, 39	23, 44, 33	55, 25, 20

Source. Ref. 45.

The figures are hypothetical and loosely based on English car sizes and
occupancies.

Table 2 lists some of the results obtained by substituting these figures and
the appropriate values for λ into Eq. 24. It is seen that, first, the roadway
space increases much more rapidly than the number of workers in the central
business district, and, second, that if all travel is by car, large town centers
require more space for roadway than for parking even if all parking is at
ground level. The second conclusion is most remarkable. In judging these
figures, one should recall that present day central business districts devote
from 15 (Tokyo) to 35 (New York) to 70% (Los Angeles) to streets and
sidewalks.

Models of this type can be used to study circumferential roadways, open
space in the middle of the city, and more complex problems. They have been
developed further for central business districts by Lam and Newell,[47] and for
satellite town commuting by Tan.[48]

III. FACILITY PLANNING AND TEMPORAL CONTROL

Viewpoint

If he were to use an engineer's jargon, the calloused politician might declare that the objectives of road network planning are (1) to determine the level of spending on transport that society is eager and willing to sustain, and (2) to allocate this flow of money in such a way that the variance among users is minimized for some appropriate dissatisfaction index. In short, spend what you can get in as fair a way as possible. All this approach requires is the ability to measure level of service in some quantitative manner so that it is possible to deduce what part of the network is weakest and therefore the most deserving candidate for the next unit of investment. To aid this sort of analysis, a wide variety of level of service criteria have been developed, with many documented in the *Highway Capacity Manual*.[49]

Before one plunges into some more complicated planning philosophy, it is worth conceding that there is much to be said in favor of the above principles. It is very difficult to decide objectively what fraction of national income should be spent on transportation, even if the arguments are based on the most innocuous economic cost–benefit criteria. Even if this fraction of income were deduced unambiguously, with full agreement by all powerful schools of economists, it would be almost impossible to apply it in a democratic society unless it coincided with the average consumer's prejudices. Furthermore, while the planner might be tempted to press for actions with the greatest economic return, the consumer will resist if the results appear to be imbalanced to the point of being grossly unfair to a vocal minority. It is for these reasons that planners often have failed in implementing economically sound plans in congested areas. On the one hand, society is reluctant to accept communal transit if it is at all able to finance private auto facilities, even if the cost of the latter is painfully large. On the other hand, because society is constrained to be at least moderately fair, it often is unwilling to uproot established communities, even when small, to implement a network improvement plan, even when the beneficiaries are numerous. Though such reactions stymie the cost–effectiveness expert, they cannot and should not be suppressed in a democratic society.

These remarks are not intended to suggest that one should quit striving for efficiency. On the contrary! But one ought to accept the idea that the objective of social planning is satisfaction, not productivity, and that a satisfied society is not always the most efficient one.

Network Performance Measures

Suppose there are several plans on hand for the road network of a given area. The differences among them may be physical, with alternative major road locations, or they may be managerial, with alternative control systems or user pricing policies. In any event, one ought to have measures which provide understandable comparisons among the alternatives. As discussed in the previous paragraph, the uses to which such measures can be put may be controversial, particularly when the planner attempts to make a guiding principle out of some raw measure of overall cost or profit.

On the whole, the measures suggested below emphasize the traveler's viewpoint. From the transportation manager's viewpoint, they do not fully describe the performance of the network. The traveler wishes to travel quickly, fairly, reliably, safely, and comfortably. The manager also wishes that his facilities are used efficiently in the sense that all investment is used with good productivity and that all usage accomplishes the designed travel purpose.

The following notation will be used:

t_{ij} = number of person trips, from zone i to zone j, in a study period such as a day or rush hour.

$T = \sum_{i=1}^{N} \sum_{j=1}^{N} t_{ij}$ = total number of trips, among the N zones of the study, in the study period.

t_{Vij} = number of vehicle trips, from zone i to zone j, in the study period.

c_{ij} = the average cost of traveling from zone i to zone j.

c_{Wij} = the cost of traveling from zone i to zone j on the minimum cost route between i and j.

Some of the more basic performance measures are the following.

MEAN TRAVEL COST PER TRIP. To achieve a *cheap* system, from the user's viewpoint, one minimizes the mean travel cost per trip, $\bar{c}$:

$$\bar{c} = \frac{1}{T} \sum_{i=1}^{N} \sum_{j=1}^{N} c_{ij} t_{ij}. \tag{25}$$

Usually c_{ij} is made equal to travel time, though it may be a more complex function of time, money, comfort, and reliability. It should be realized that

$$\bar{c}_{Wij} = \frac{1}{T} \sum_{i=1}^{N} \sum_{j=1}^{N} c_{Wij} t_{ij} \tag{26}$$

is not generally equal to $\bar{c}$ as obtained by minimizing Eq. 25. In Eq. 26, each traveler takes "the worm's eye view" and selects a minimum cost route for himself alone, regardless of its effects on other drivers. The minimization of Eq. 25 yields a minimum overall cost for the traveling society. To achieve this usually requires that some duress—economic, psychological, or legal—be put on the traveler.

The reason that the two measures differ is that travel time on a route increases with the number of travelers using that route. Thus when virtually all drivers take the fastest route between two points, everyone doing so travels more slowly than he would if some fraction of the group were to take a slower route. Naturally no one volunteers to take a slower route just so everyone else can go a bit faster, even if the sum of all travel times would thus be reduced.[50]

Another measure sometimes used is

$$\bar{c}_V = \frac{1}{T} \sum_{i=1}^{N} \sum_{j=1}^{N} c_{ij} t_{Vij}. \tag{27}$$

This gives the average vehicle travel cost in the system, provided c_{ij} is made the trip cost per vehicle, rather than per person. Though Eq. 27 is a better measure than Eq. 25 for system occupancy and therefore has meaning for the system manager, it suppresses a major aspect of the customer's experiences. If Eq. 27 were minimized, one would give no preference to buses, which each carries many travelers. On the other hand, minimization of Eq. 25 implies such a preference, usually desirable from the customer viewpoint.

VARIANCE OF TRAVEL COST AMONG CUSTOMERS.   To achieve a *fair* system, one minimizes the variance of travel cost among customers,

$$\text{var}(c) = \frac{1}{T} \sum_{i=1}^{N} \sum_{j=1}^{N} (c_{ij} - \bar{c})^2 t_{ij}. \tag{28}$$

MEAN COEFFICIENT OF VARIATION FOR INDIVIDUAL'S TRAVEL COSTS.   To achieve a *reliable* system, one minimizes the mean coefficient of variation for the individual's travel costs,

$$\bar{V} = \frac{1}{T} \sum_{i=1}^{N} \sum_{j=1}^{N} V_{ij} t_{ij}, \tag{29}$$

where $V_{ij} = [\text{var}(c_{ij})]/(c_{ij})$ and $\text{var}(c_{ij})$ is the variance in travel cost on route ij among travelers and over a number of equivalent analysis periods.

CHANGE IN MEAN TRAVEL TIME WITH A CHANGE IN THE MEAN NUMBER OF TRAVELERS.   A system *saturation* measure is provided by the derivative of mean travel cost with respect to the mean number of travelers. The change in the number of travelers, dT, is allocated among all routes on the basis of how trips are distributed, so $dt_{ij} = (t_{ij}/T) dT$. The saturation measure S is

$$S = \left(\frac{d\bar{c}}{dT} \right), \tag{30}$$

with dT allocated as described above.

MEAN VELOCITY GRADIENT. To achieve a *comfortable* and *safe* system, one minimizes the average mean velocity gradient $\bar{G}$, where

$$G_{ij} = \frac{1}{X\bar{v}} \int_0^X [a(y)]^2 \, dy, \tag{31}$$

$$\bar{G} = \frac{1}{T} \sum_{i=1}^N \sum_{j=1}^N G_{ij} t_{ij}, \tag{32}$$

and X, $a(x)$, $\bar{v}$ are the time duration, acceleration as a function of time, and average velocity for a trip between zones i and j. Since G is large if movement is uneven, and there is much acceleration and braking, the minimization of G improves comfort and safety.[51]

One trouble with simple performance measures, such as the examples just given, is that only one measure can be optimized with certainty in any one system. For example, there is no reason why a system that minimizes cost will also minimize the variance in day-to-day costs. One has the options of (1) optimizing just one performance measure, (2) optimizing some composite utility function, or (3) optimizing one or a combination performance measure subject to a series of constraints in the form of preset minimum standards for service. It is approach (3) that usually is taken. Thus one might seek to minimize costs subject to minimum standards of fairness, reliability, and safety. Of course, overall costs include construction and maintenance expenses, as well as the network-route-dependent traveler costs with which we are concerned here.

Once some performance or utility measure has been decided on, alternative networks can be compared theoretically by assigning traffic to each of their graphs and then comparing the appropriate measures. However, if the competing networks show dramatically different travel times or patterns, it is necessary to iterate the overall prediction methodology by reconsidering trip generation and distribution until these are fully consistent with the proposed networks. Otherwise, there may be substantial errors because of transport-system-induced changes in land use and travel patterns.

Facility Planning

Physical facilities, such as roads or bridges, usually are decided on by first discerning a need, then postulating a small number of alternative projects, and finally deciding on one of these alternatives on the basis of a cost–effectiveness or other utility measure. One is not assured of finding the best solution because the small number considered may not include the optimum one.

In some circumstances it may be possible to search systematically for the best among all possible solutions. One problem area where such systematic search has been proposed and, to some extent, tested, is the selection of a route for a new road. The simplest version of this problem is the siting of a new bridge or tunnel. Here there is only one variable to be tested, distance along the barrier. Here there is no great difficulty in making a series of assignments, for each of several crossing locations, in a systematic manner that will yield the optimum point.

A substantially more difficult problem is the siting of a road. One seeks the best route function $y = S(x)$ in a two-dimensional space (x, y) between origin and destination. It may be that one can approximate the cost of the road, entirely in terms of the route function $S(x)$ and its derivative $dS(x)/dx$, in such a way that an infinitesimal stretch of road at (x, y) has a cost $f[x, S(x), dS(x)/dx] \, dx$. Then, if the cost function f were known for all (x, y), one could calculate the cost of any route $S(x)$ by evaluating the integral for cost J:

$$J = \int_{x_0}^{x_1} f\left(x, S(x), \frac{dS(x)}{dx}\right) dx, \tag{33}$$

where x_0 and x_1 are the end points of the road. If the cost function can be exhibited in the form of Eq. 33, the cost J is minimized by that route $S(x)$ that satisfies the Euler–Lagrange equation

$$\frac{\partial f}{\partial S} = \frac{d}{dx}\left(\frac{\partial f}{\partial (dS(x)/dx)}\right) = 0. \tag{34}$$

Reference 52 gives a thorough explanation, with examples, of the application of this variational calculus approach to road siting.

Another problem amenable to systematic solution is the siting of interchanges on limited access expressways. This problem is a close analog to the railroad station spacing problem, which latter was analyzed in connection with the BART system for San Francisco area commuting.[53] The interchange-dependent part of the road cost can be developed as a function of the number of interchanges, the origins and destinations of the travelers, and the unit costs of travel both on and off the expressway. The derivative of this cost function with respect to the number of interchanges, set equal to zero, gives the minimum cost number of interchanges and, from this number, the optimum interchange spacing.

Facility Management

The main efforts of research in road network management have been directed at the optimum timing of one- and two-dimensional arrays of signal lights. These problems are discussed in Chapter 3 of this book. A network assignment model can be used for these signalization studies and for other time-dependent control systems only if it itself is a fully time-dependent formulation. Though practical to program for a digital computer,[54] such a formulation does not avoid immense difficulties in model calibration and in the cost of the required computer time. As a result, most facility management studies have relied on highly idealized, special purpose models, wherein no effort is made to engage in precise individual vehicle assignment.

The following discussion is limited to applications of the basic principle that congestion is costly and should be minimized. Though the principle is obvious, some of its applications lead to perhaps surprising results. The usual cause of unnecessary congestion is network occupancy by vehicles that are not accomplishing anything by their occupancy and that wish to end it as expeditiously as possible. Some examples follow.

VEHICLES TRAVELING IN SEARCH OF PARKING SPACE.   A moving vehicle takes more space than a parked one. If the vehicle moves about the network for a period of time before parking and if it does not give up until it parks, as is the situation for commercial vehicles in a crowded central business district, then it is obvious that much more of the available space–time inventory is given to that vehicle by the network than would be the case if the vehicle were able to park at once. Thus it may be false to claim that very little street parking space should be provided in congested areas. Even though a re-duction in the proportion of space devoted to parking does increase street capacity, the added capacity may be oversubscribed by the increase in the number of vehicles cruising about while looking for parking. There is an optimum mix, between moving and parking space in such a situation, which can be calculated with the aid of queueing theory.[50]

The same sort of problem arises when cars search along free or cheap street curbs for parking before they are prepared to enter a more expensive off-street parking facility. The obvious solution here is to have all available parking space priced at the same level. Further, this price should be high enough to ensure that parking is available almost all the time. If this is done, a vehicle that has accepted the cost and has come to park will not waste or extend its trip, and the network facilities therefor, without accomplishing promptly the parking at the trip end, which is required to justify the overall cost of the trip.

VEHICLES PARKED ON HIGH-COST FACILITIES.   If the density of vehicles on freeways, bridges, tunnels, or other high-cost facilities is too high, their throughput declines from optimum and their high cost is not fully exploited. It is better to hold vehicles back, or transfer them to alternate routes, and thus maintain full productivity. The localized problem of metering traffic onto a high-cost facility, for example at an expressway ramp, is a queueing problem, discussed in Chapter 3 of this book. In addition, Ref. 55 gives the results of some recent practical work on ramp metering. The broader network problem of allocating traffic among expressway entrances or among alternate routes can be treated with a capacity restraint assignment model wherein routing is not left entirely to the discretion of the drivers.

THE COLLECTION OF TOLLS TO REDUCE CONGESTION. Since excessive occupancy of the network produces delay, and delay is costly, it may be worthwhile to control occupancy by a toll collection system. In the absence of control, *congestion pricing* prevails. The drivers who use the network are willing to pay the time costs of congestion. Alternatively, one might discourage some of these drivers by instituting a toll charge whose purpose would be to limit traffic to the level at which one additional vehicle's benefit in using the system would just equal all the other vehicles' loss due to the increased travel time imposed by the additional vehicle. This level of traffic would be the most efficient from an economic viewpoint. The approach does raise serious questions about the fairness of price rationing. However, this sort of rationing already prevails in the distribution of most consumer resources. In the street network it can be seen to work in areas where parking is limited and expensive. Reference 56 presents a powerful case for money as against congestion pricing.

References

1. L. R. Ford and D. R. Fulkerson, *Flow in Networks*, Princeton: Princeton University Press, 1962.
2. R. B. Potts and R. M. Oliver, *Flows in Transportation Networks*, New York: Academic, 1972.
3. F. V. Harmann, *Population Forecasting Methods*, Bureau of Public Roads, Washington, D.C., 1964.
4. F. V. Harmann, "Population Estimates," *Current Population Reports*, Series P-25, No. 110, Washington, D.C.: Bureau of the Census, 1955.
5. Portland City Planning Commission, *Portland's Economic Prospects*, Portland, 1957.
6. Denver Planning Office, *Working Denver*, Denver, 1953.
7. B. R. Berman, B. Chinitz, and E. M. Hoover, *Technical Supplement to the New York Metropolitan Regional Study*, Cambridge: Harvard University Press, 1961.
8. S. H. Putnam, "Analytic Models for Implementing the Economic Impact Studies for the Northeast Corridor Transportation Project," 30th National Meeting, Operations Research Society of America, 1966.
9. C. N. Swerdloff and J. R. Stowers, "A Test of Some First Generation Residential Land Use Models," 45th Highway Res. Board Meeting, Washington, 1966.

10. J. R. Hamburg and R. H. Sharkey, "Land Use Forecast," *Chicago Area Transportation Study*, Chicago, 1961, 3.2.6.10.

11. J. D. Herbert and B. H. Stevens, "A Model for the Distribution of Residential Activity in Urban Areas," *J. Regional Sci. Assoc.*, Fall, 1960.

12. B. Harris, "Linear Programming and the Projection of Land Users," *Penn-Jersey Transportation Study*, PJ Paper No. 20, 1962.

13. W. G. Hanson, "Land Use Forecasting for Transportation Planning," *HRB Bull.*, **253**, 145–151 (1960).

14. S. A. Stouffer, "Intervening Opportunities," *Amer. Sociological Rev.*, **5**, 845–867 (1940).

15. M. Schneider, *Chicago Area Transportation Study*, Vol. 2, 1960.

16. G. G. Sherratt, "A Model for General Urban Growth," *Management Sciences—Models and Techniques*, New York: Pergamon, 1960, Vol. 2, pp. 147–159.

17. W. Helly, "A Time-Dependent Intervening Opportunity Land Use Model," *Socio-Economic Planning Sci.*, **3**, 65–73 (1969).

18. H. D. Deutschman, "Auto Ownership Revisited," *Highway Res. Rec.*, **205**, 31–49 (1967).

19. *Highway Res. Rec.*, particularly Nos. **114** (1966), **165** (1967), **191** (1967), **205** (1967), and **250** (1968). These and expected later issues on "Origin and Destination Characteristics" form a journal of practical experience with trip generation, distribution, and assignment models.

20. T. J. Fratar, "Forecasting Distribution of Interzonal Vehicular Trips by Successive Approximations," *Proc. Highway Res. Board*, **33**, 376–385 (1954).

21. G. E. Brokke and W. L. Mertz, "Evaluating Trip Forecasting Methods with an Electronic Computer," *Highway Res. Board Bull.*, **203**, 52–75 (1958).

22. A. M. Voorhees, "A General Theory of Traffic Movement," *Proc. Inst. Traffic Engrs.*, 46–56 (1955).

23. Chicago Area Transportation Study, *Final Report*, 1960, Vol. II; also E. R. Ruiter, "Improvements in Understanding, Calibrating, and Applying the Opportunity Model," *Highway Res. Rec.*, **165**, 1–21 (1967).

24. R. W. Whitaker and K. E. West, "The Intervening Opportunity Model: A Theoretical Consideration," *Highway Res. Rec.*, **250**, 1–8 (1968).

25. R. B. Potts, *Traffic Flow in Networks*, Summer School on Traffic Theory, Purdue University, 1967; see also Ref. 2, pp. 121–135.

26. A. G. Wilson, "A Statistical Theory of Spatial Distribution Models," *Transport. Res.*, **1**, 253–269 (1967).

27. M. J. Beckmann and T. F. Golob, "A Critique of Entropy and Gravity in Travel Forecasting," Ref. 35, pp. 109–117.

28. T. W. Anderson, *Introduction to Multivariate Statistical Analysis*, New York: Wiley, 1958.

29. P. G. Hoel, *Introduction to Mathematical Statistics*, New York: Wiley, 1947, p. 121.

30. T. Deen, W. Mertz, and N. Irwin, "Application of a Modal Split Model to Travel Estimates for the Washington Area," *Highway Res. Board Rec.*, **38**, (1963).

31. S. L. Warner, *Stochastic Choice of Mode in Urban Travel*, Northwestern University Press, Evanston, 1962.
32. P. R. Stopher, "A Probability Model of Travel Mode Choice for the Work Journey," *48th Annual Meeting of the Highway Res. Board*, Washington, 1969.
33. H. W. Bevis, "A Probabilistic Model for Highway Traffic Assignment," *49th Annual Meeting of the Highway Res. Board*, Washington, 1970.
34. R. H. Pratt, "A Utilitarian Theory of Travel Mode Choice," *49th Annual Meeting of the Highway Res. Board*, Washington, 1970.
35. C. G. B. Mitchell and J. M. Clark, "Modal Choice in Urban Areas," *Traffic Flow and Transportation*, edited by Gordon F. Newell, New York: Elsevier, 1972, pp. 71–85.
36. U.S. Department of Commerce, Bureau of Public Roads, *Traffic Assignment Manual*, Washington, 1964.
37. T. Hoshino, "Theory of Traffic Assignment to a Road Network," Ref. 35, pp. 195–214.
38. F. S. Hillier and G. J. Lieberman, *Introduction to Operations Research*, San Francisco: Holden-Day, 1967, pp. 218–222; see also Ref. 16, pp. 130–134.
39. E. F. Moore, "The Shortest Path Through a Maze," *Proc. Intern. Symp. on the Theory of Switching*, Cambridge, Mass.: Harvard University Press, 1957.
40. A. Shimbel, "Applications of Matrix Algebra to Communications Nets," *Bull. Math. Biophysics*, **13** (1951).
41. P. Loubal, "A Network Evaluation Procedure," *Highway Res. Rec.*, **205**, 96–109 (1967).
42. R. R. Snell et al., "Traffic Assignment with a Non-Linear Travel Time Function," *Transport. Sci.*, **2**, No. 2, 146–159 (1968).
43. T. F. Humphrey, "A Report on the Accuracy of Traffic Assignment When Using Capacity Restraint," *Highway Res. Rec.*, **191**, 53–75 (1967).
44. J. D. Carroll, "A Method of Traffic Assignment to an Urban Network," *Highway Res. Bull.*, **224**, 64–71 (1959).
45. B. V. Martin, *A Computer Program for Traffic Assignment Research*, Cambridge, Mass: Department of Civil Engineering, M.I.T., 1964.
46. R. J. Smeed, *The Traffic Problem in Towns*, Manchester: Manchester Statistical Society, 1961.
47. T. Lam and G. Newell, "Flow Dependent Traffic Assignment in a Circular City," *Transport. Sci.*, **1**, No. 4, 318–361 (1967).
48. T. Tan, "Mathematical Model for Computer Traffic in Satellite Towns," *Transport. Sci.*, **1**, No. 1, 6–23 (1967).
49. U.S. Department of Commerce, Bureau of Public Roads, *Highway Capacity Manual*, 1965.
50. W. Helly, "Efficiency in Road Traffic Flow," *Proc. 2nd Intern. Symp. on Theory of Traffic Flow*, Paris: O.E.C.D., 1965, pp. 264–276.
51. W. Helly and P. Baker, "Acceleration Noise in a Congested Signalized Environment," in *Vehicular Traffic Science*, edited by L. C. Edie, New York: Elsevier, 1967, pp. 56–61.

52. P. Friedrich, "Die Variationsrechnung als Planungsverfahren der Stadt und Landesplenung," *Schriften der Akademie fur Raumforschung und Landesplanung*, Bremen-Horn: Walter Dorn, 1956, Vol. 32; see also B. C. Kahan, *Res. Note RN/4037/BCK*, Road Research Laboratory, Harmondsworth, Middlesex, England, 1961.

53. V. Vuchic and G. F. Newell, "Rapid Transit Interstation Spacings for Minimum Travel Time," *Transport. Sci.*, **2**, No. 4, 303–339 (1968).

54. R. C. Boehne and R. P. Finkelstein, *Rational Approach to Urban Highway Congestion*, Stanford, Calif.: Stanford Research Institute, 1967.

55. Texas Transportation Institute, Texas A & M University, *Gap Acceptance and Traffic Interaction in the Freeway Merging Process*, 1969.

56. G. Roth, *Paying for the Roads*, Harmondsworth, Middlesex, England: Penguin, 1967.

Index

Acceleration, measurement of, 16
Acceleration lane, merge from, 127
Acceleration noise, definition of, 92
  measurement of, 93
Accessibility model, 256
Accidents, and car-following, 101
Adaptive control, of many intersections, 204
Adjustment, process in driving, 35-36
Allocation, of roadway use, 225
  of traffic facilities, 229
Arrival, cumulative, 210
  process of, 143
Arrival process, compound Poisson, 151
  at intersection, 143
  Poisson, 150
Assignment, all-or-nothing, 269, 274
  diversion, 268
  incremental, 275
  multiple route, 275
  semi-time-dependent, 246
  time-dependent, 246
  time-independent, 246
  of traffic, 246, 269-279
  of trips, 248
Automatic highways, 100

Bandwidth, design for maximum, 180-182
  of a progression design, 184
Bang-bang control, 213, 215
Bellman's principle of optimality, 198, 202
Boltzmann theory, 7, 35
Bottlenecks, creation of, 5, 62
  fixed, 62
  moving, 67
  queueing at, 137
Braking distance, 13
Bus-following, experiments, 98
  response-time in, 98

California Vehicle Code, 13, 70
Capacity restraint, 224, 270, 274
Car-following, and accidents, 101
  automatic control in, 100
  buses, 98
  experiments, 90
  lag-time in, 91, 92
  memory in, 97

models of, 86-102
non-linear models of, 95
parameters in, 91, 97
speed-spacing models, 98
stability in, 89
steady-state, 96
stimulus-response in, 87
theories of, 6, 86-102
CDS diagram, 222
Chicago, Eisenhower expressway, 230
Central business district, 284
Centroids, of network, 242
Clustering of vehicles, 44
Combination method, 192
Computer-control, of San Jose system, 189
  of traffic signals, 177, 189
  of tunnels, 233
Concentration, jam-, 43
  prediction of, 66
Condensation of network, 195
Congested transportation systems, 224-229
Congestion pricing, 285
Conservation, definition of, 8
  of trips, 265
  of vehicles, 6
Control, bang-bang, 213, 215
  of critical traffic links, 229
  domain of, 225
  of freeways, 229, 230
  major loop, 180
  minor loop, 180
  region of, 211
  temporal, of facilities, 279
  of tunnels, 229
Cordon of a network, 244
Critical traffic links, control of, 229
Cumulative arrival, 210
Cumulative service, 210
Cycle time, optimum, 160, 214
  for oversaturated intersection, 214

Delay, average, pervehicle, 178
  effect of cycle time on, 156
  for general arrival process, 148
  at intersection, 113, 123, 149
  of left-turners, 141
  maximum individual, 178
  mean, 121

for Poisson input, 146
probability density of, 117
to a queue of cars, 141
rush-hour, 159
of a single car, 117
at a stop sign, 121
at a traffic signal, 142, 145
Delay time, first and second moments of, 119
mean and variance of, 119
at multilane crossings, 125
optimization of, 159
Density, measurement of, 234
oscillations, 48
Density-saturation gradient method, 256
Departure process, 144
Detroit, John Lodge expressway, 230
Diagram, flow-concentration, 54
of time sequence, 18
space-time, 54, 181
Discretization, of a control theory problem, 226
Discriminant function, 267
Distribution, counting, 21, 25
displaced exponential, 24
fitted to data, 23
hyperlang, 27
negative exponential, 24, 111
normal, 23, 31
of free speed, 7
of headways, 21, 112
of traffic variables, 20
of trips, 242, 248, 260
peaking factor, 25
Poisson, 21
Disutility, of a trip, 267
Diversion assignment, 268
Driver, impatient, 120
PIEV, 4
reaction time of, 4, 82
response of, 4
Driver-vehicle, response time, 82
Driving, California Code, 70
Domain, control, 225
space-time, 10, 16

Economic forecasts, 251
Entropy minimization, 265
Equations, of continuity, 57
of motion, 70
of time-dependent flow and concentration, 75
of traffic state, 57
EQUISAT design, 205
Experiments, bus-following, 98
car-following, 90
controlling concentration, 82
flow-concentration, 72
flow-speed, 81
with gaps in streams, 80
macroscopic in tunnels, 76
Expressway, control of, 230
Eisenhower, 230

John Lodge, 230
spillback from exit ramp of, 218

Facility, management, 283
planning, 279, 282
Flow, definition of, 8
nonstationary, 127
Flow-concentration, contour maps, 64
curve, 54, 62
experimental data, 72
linear curve, 70
logarithmic curve, 71
models, relationships, 73
parabolic curve, 70
triangular curve, 82
Forecasts, economic, 252
land use, 255
population, 251
Freeways, control of, 230
density measurement, 236
Friction between zones, 263

Gamma distribution, of headways, 28
Gap acceptance, in expressway control, 231
function of, 114-117
generalized function of, 121
for multilane highways, 124
parameter estimators of, 116
process of, 135
step function, 117, 120
at stop-controlled intersection, 116
Gaps, between queues, 113
blocks and, 113, 124, 140
critical, 116, 123, 138, 139
Generation of trips, 242, 248, 250-260
Glasgow experiments, 199, 205
Gravity model, 256, 263
Growth factor, average, 262
Fratar model, 262
models, 262

Headway, definition of, 11, 111
distribution of, 21, 28
mean, 111
successive properties of, 112
Headway distribution, Erlang, 28
fitted to data, 29
gamma, 28
for light traffic, 121
log-normal, 28
mixed, 28
semi-random, 28
Heuristic control algorithms, 199-209
Highways, automatic, 100
Hill-climbing procedure, in TRANSYT method, 192
Hudson river, tunnels under, 229
Hydrodanamic theory, introduction to, 52
quantitative models, 69

IBM, joint study with city of San Jose, 189
joint study with Port of New York Authority, 233

1800 traffic control system, 189
Index of dispersion, for arrival process, 143, 144
    for Poisson process, 143
Input ramp, merging from, 115
Input-output model, 252
Instability, in car-following, 89
    in tunnels, 69
Interaction process in driving, 35
Intersection, controlled, 142
    delay optimization for, 143
    oversaturated, 209, 226
    uncontrolled, 142
Intersection capacity, effect of critical gaps, 123
Intervening opportunity models, 256, 264

Kalman filtering, 235
Kinetic theory of gas, analogy to traffic, 7
    Boltzmann theory, 7
Kirchhoff's laws, analogy to, 198

Labeling strategy for networks, 246
Lag acceptance, at stop-controlled intersections, 116
Lag-time in car-following, 91, 92
Land use accounting model, 276
Land use forecasts, 255
Lane changes, model of, 49
Left-turners, delay of, 141
Length, measurement of, 16
Leontief, input-output model, 252
Light traffic approximation, 122
Limit cycle of traffic light, 201
Lincoln tunnel, control of, 233
    measurement of density at, 234
Link importance factor, 197
Link of network, 244
Log-normal distribution of headways, 28

Macrocontrol, 180
Macroscopic models, 51-86
Markov chain, embedded, used for pedestrian queueing, 128
    used for vehicular queueing, 138
Measurement, of acceleration noise, 93
    of vehicle acceleration, 16
    of vehicle length, 16
    of vehicle spacing, 16
    of vehicle speed, 16
Memory, in car-following, 97
Merges, delayed, 117
    multiple, 116
    onto a freeway, 115
    onto single lane, 117
    single-vehicle, 116
Merging, and queueing, 114
    onto a freeway, 115
    from acceleration lane, 115
Microcontrol, 180
Microscopic models, 86-102
Microscopic theory, application of, 99
Minimum cost, route algorithm, 270

Minimum entropy model, 265
Minimum time tree, 274
Modal split, 266
Models of car-following, 86
Monte Carlo game, in SIGOP design, 197
Multilane highway, crossing of, 124
    gap acceptance for, 124
Multiple route assignment, 275

Networks, centroids of, 242
    condensation of, 195
    connected, 244
    definitions, 242
    flow in, 245
    graph of, 244
    irreducible, 195
    links of, 244
    nodes of, 244
    occupancy, 248
    performance measures, 279
    simulation of, 250
    sinks of, 244
    sources of, 244
    spider-web, 242
    store-and-forward, 224
    zones of, 242
New York City, tunnels into, 229
Nodes of networks, 244

Pedestrian, crossing, 124, 132
    crossing of two lanes, 132
    expected delay of, 133
    flyovers, 134
    island, 132
    Poisson arrival of, 128, 132
    push-button crossings, 132
    queueing, 127
    subways, 134
Platooning, 30
Platoons, definition of, 26
    diffusion of, 30, 189
    dispersion of, 189, 191
    movement through traffic lights, 188
    starting characteristics, 102, 188
Poisson distribution, 21, 23
Policy improvement, 205, 217
Pontryagin's maximum principle, 214
Population forecasts, 251
Port of New York Authority, joint study with IBM, 233
Probit, definition of, 116
Progression design, efficiency of, 182
    first appearance, 177
    properties of, 182
    reverse, 188
PR-system, 177

Queueing, bulk, 134
    of pedestrians, 130
    vehicular, 134
Queueing theory, approach to traffic, 44
    influence of passing, 46
Queues, behind truck, 67

definition of, 123
gaps between successive, 113
of pedestrians, 130
random, 123
splitting of, 138
travelling, 112, 123

Ramp, closure of entrance, 231
  spillback from exit, 218
Ratio method for forecasts, 252
Reflecting barriers, in traffic light operation, 200
Regeneration point, 124
Relaxation process in driving, 35
Renewal hypothesis for headways, 29
Response time, of buses, 98
  of driver, 14
  of driver-vehicle complex, 82
Reversible lanes, 221
Revolving-door effect, 229
Roadway, bottlenecks, 5
  capacity, 5
  parameters, 4
Robustness, of delay calculations, 123
  of models of delay, 116
Route algorithm, minimum cost, 270
Route assignment, 225
Route control, benefits from, 206
  by directional instructions, 206
  near critical intersections, 207

San Jose system, 188, 199
Saturation flow, algorithm, 201
  definition of, 191
Saturation measure, 281
Screen line, 244
Semi-Markov models, 127
Sensitivity coefficient, for buses, 98
  for cars, 87
Shockwaves, acceleration, 69
  deceleration, 69
SIGOP design, 196
Signal discipline, 144
Simulation of multilane crossings, 127
Siting, of a road, 283
  of interchanges, 283
Space-time domain, 10, 16
Spacing, California Vehicle Code, 13
  definition of, 11
  as a function of speed, 12
  measurement of, 16
Speed, definition of, 8
  harmonic mean, 8
  mean free, 43
  measurement of, 16
  space mean, 8
  versus concentration, 43, 57
Speed distribution, beta, 33
  Boltzmann-type, 34
  desired, 31, 32
  experimental, 39, 40
  exponential, 33
  fitted to data, 37

free, 31
on freeways, 31, 37
gamma, 33
on George Washington Bridge, 38
mixed, 33, 34
theoretical, 39, 40
on urban streets, 31
Spider-web network, 242
Spill-back from exit ramp, 218
Split of traffic light, 178
Stability, in car-following, 89
  of multilane traffic, 49
Stacking of aircraft, 224
Starting wave, description of, 59
  influence on synchronization design, 188, 190
State variables, 214, 225
Statistical mechanics, model for traffic, 35
Steady-state, in car-following, 96
Ste-down method, for forecasts, 252
Stimulus-response, in car-following, 87
Storage capability of arcs, 224
Store-and-forward networks, 224, 231
Stop sign, 114
Switch-over time, 213
Synchronization of traffic lights, definition of, 176
  by discrete maximum principle, 198
  by dynamic programming, 198
  for maximum bandwidth, 180, 182
  by mixed-integer linear programming, 185
  problem of transition, 206
  San Jose design, 188
  schemes of, 180-199
  by SIGOP design, 196
  by TRANSYT method, 188, 191

Temporal control, of facilities, 279
Through-band, design for maximum, 180, 182
Throughput of a traffic system, 178
Time-lag, in car-following, 87
  for lane-changing, 49
Time-sequence, diagram of, 18
Traffic, control of, 175-239
  generation, distribution, and assignment, 241-288
  heavy, 51
  light, 42
  moderate, multilane, 48
  moderate, single-lane, 44
  sinks of, 189
  sources of, 189
  state changes of, 56
Traffic-actuated signals, delay at, 162-169
  evolution of, 177, 199
  fully actuated, 199
Traffic characteristics, 3
Traffic concentration, 3
Traffic control, criteria of performance, 178
  evolution of, 177
  heuristic schemes for, 179

in tunnels, 80, 233
    objective function, 178
    objectives of, 178
    open-loop, 176
    principles of operation, 178
Traffic density, 3
Traffic dynamics, macroscopic, 51
Traffic flow, analogies for, 5
    definition of, 8, 9
    different regimes, 81
    free-flowing, 26
    hydraulic analogy, 8
    maximum, 13, 14
    parameters, 3
    restrained, 26
    starting from signal, 83
    theories, 2
    variables, 3, 20
Traffic flow theories, approaches to, 5
    macroscopic, 6, 51-86
    microscopic, 6, 86-102
    physical analogies, 7
    statistical, 6
    two-regime fit, 77
Traffic intensity, heavy, 4
    light, 4
    moderate, 4
Traffic Research Corporation, 196
Traffic signal, delay at, 141, 151
    fixed-cycle, 141
    limit cycle of, 201
    offset of, 178
    optimization of, 159
    overflow at, 151
    split of, 178
    the first, 176
Traffic stream, discrete structure, 25
Traffic variables, distribution of, 20
Trajectories, observation of, 63
    of vehicles, 55
    of waves, 55
TRANSYT method, 188, 191
Trap, of observations, 15

Travel cost, mean, pertrip, 280
    variance, among customers, 281
Travel time, change with number of travel-
        ers, 281
    in urban environment, 83
Tree, network, 244
Trip, assignment, 242, 248
    distribution, 242, 248
    disutility of, 267
    generation, 242, 248, 250, 259
    table, 244
Tunnel, bottlenecks in, 76
    control of traffic in, 80, 233
    degradation of efficiency, 229
    instability in, 69

Unit extension, of traffic light phase,
        199
U. S. Bureau of Public Roads, 196

Variables, control, 214, 225
    state, 214, 225
    trip predictor, 251
Vehicle-actuated signal, delay at, 162-169
Vehicle parameters, measurement, (accelera-
        tion, length, spacing, and speed), 16
Vehicular queueing, 134
Velocity gradient, mean, 282

Waves, celerity of, 54
    flow, 56
    propagation of, 52
    shock, 52, 53, 56
    starting, 59, 188, 190
    stopping, 59
    speed of, 6, 82
    trajectories of, 55

Yield sign, delay at, 119

Zebra crossings, definition of, 132
    economic benefits from, 134
Zones of network, 242